Extremal Combinatorial Problems and Their Applications

Mathematics and Its Applications

Managing Editor:

M. HAZEWINKEL

Centre for Mathematics and Computer Science, Amsterdam, The Netherlands

Volume 335

Extremal Combinatorial Problems and Their Applications

by

Boris S. Stechkin
Steklov Mathematical Institute,
Russian Academy of Sciences,
Moscow, Russia

and

Valeriy I. Baranov
Korvus Company,
Moscow, Russia

SPRINGER-SCIENCE+BUSINESS MEDIA, B.V.

A C.I.P. Catalogue record for this book is available from the Library of Congress.

ISBN 978-94-017-4122-4   ISBN 978-0-585-29602-9 (eBook)
DOI 10.1007/978-0-585-29602-9

Printed on acid-free paper

To Our Mothers

Table of Contents

Preface to the English edition

The English version of this originally Russian book has been substantially supplemented by new material, and is almost 50 percent larger than the original. Part of the new material has been prepared in cooperation with colleagues. We express our deep gratitude to all of them: A. Klimov, A. Kostochka, I. Kan, I. Rival, V. Shmatkov, K. Rybnikov, A. Malykh, S. Sal'nikov, N. Sauer, A. Sidorenko, J. McIntosh, W. Kocay, C. Dodson, S. Radziszowski, V. Rodl, R. Wilson, and Gy. Katona.

In particular, Jack McIntosh offered the use of the word 'packability' for our very new Russian term 'vlozhimost''.

Our special thanks go to the translator. Having made attempts ourselves to translate combinatorial literature into Russian, we know how difficult it is to render correctly thoughts sometimes expressed in ponderous language or overloaded with meaning. However, we believe that even this paragraph has been successfully translated.

We thank KLUWER Acad. Publ., who ventured to undertake this step, although we think that the risk has been substantially reduced by the superb coordination of the entire work done by Margaret Deignan, to whom we express our deepest gratitude.

Preparation of the English version of the book was partially supported for the second author by the Russian Foundation for fundamental research with its Algebraic Combinatorics grant $\sharp 93 - 011 - 1442$.

Introduction to the English edition

As far as we know, this is the first Russian book on general combinatorics that has been translated into English. In recent decades, the reverse process has occurred: most Western monographs, conference proceedings, and some collections of papers on combinatorics have been translated into Russian and published in large printruns.

In the post-War period, a vigorous development of combinatorial research has taken place in Russia on a two-fold basis: translated sources together with Russian books, conference proceedings and articles, and, finally, a journal specifically devoted to combinatorics. Russian combinatorists have been more fully informed than their Western colleagues. When reading our book one may get the feeling that we have devoted insufficient attention to foreign results. The Russian reader will know better: we exactly know its information sources and, candidly, have not feared the appearance of this translation. The book was written for Russian readers, with Russian objectives.

One such objective was to attract young people to the thematics of extremal problems and to combinatorics as a field of study. Therefore, in part, the book has the features of both a textbook and a reference handbook and can be read by mathematics students and beginning engineers. We have happily seen this objective realized in that the work of one of these students is presented in this English version by advances on the Frobenius problem.

Another objective is shown in our attempt to extend the extremal approach to solving a large class of problems including some previously regarded as exclusively algorithmic. Unfortunately, the problem '$\mathcal{P} = \mathcal{NP}$' has, from time to time, baffled not just theorists.

Related to this very objective is a third one (chronologically, the first): to broaden the freedom of choice of theoretical bases for modeling real phenomena in order to completely solve practical problems.

The real phenomenon which prompted this whole project is this: if a large number of problems (say, 108) are simultaneously solved by a computer, 'crowding' (fragmentation) of its memory occurs, sharply increasing both the total time and the separate time for solving each problem. This is sometimes of essential and even fundamental importance, for example, in the detection and servicing (destruction) of a set of fast-flying targets. And if their fly-up time (say, to Moscow) is five to eight minutes, then every computer second gained turns into entirely concrete reality.

The method is sufficiently universal, since every computer has a memory—even the abacus, which, to this day, remains an unsurpassed configuration: it is simultaneously a carrier of memory, a processor, and a display, albeit necessarily with a person present.

Moscow
27 January 1993

V.I. Baranov
B.S. Stechkin

Preface

This book is the result of the close cooperation between an engineer and a mathematician to develop methods for solving problems arising in the creation of automatic control systems. The main result of this cooperation is a combinatorial model, packability of number partitions, expounded in this book.

A study of packability of number partitions preceded the analysis of a number of practical problems arising in the design of efficient methods for controlling computer memory allocation, in the development of methods for the analysis of programming techniques for automatic control systems, etc. The choice of combinatorial methods was predetermined by the creation of the new, practically important, theme of extremal combinatorial problems on packability of number partitions. This combinatorial direction turned out to be useful not only in the formalization and solution of a number of problems in engineering, but also in solving a class of extremal problems on graphs.

The aim of this book is to let engineers and mathematicians become acquainted with the methods developed by the authors for solving a number of applied and mathematical problems. The material of the book is divided into 5 chapters.

Chapter 1 is a short guide to the necessary combinatorial notions. In particular, next to all elementary combinatorial schemes it explains a listing scheme developed by the authors, enabling us to unify the simplest combinatorial schemes.

Chapter 2 contains the basic mathematical results in the study of the packability of number partitions, and presents the, up till now, most complete summary of results in this direction. As illustration of the applicability of these results we note their relation to old weighing problems and other problem statements. We give as exercises some problems and assertions concerning the packability of number partitions.

Chapter 3 is devoted to getting acquainted with extremal problems on graphs and systems of sets. We show their relation with results on the packability of number partitions.

In Chapter 4 we study certain extremal problems in geometry, and apply the results of having solved them.

In Chapter 5 we give methods for using the results of solving extremal combinatorial problems concerning the packability of number partitions in the design of ACS. Here we give combinatorial models for studying processes to control the design of ACS and computer memory allocation. We demonstrate the applicability of the packability theorem for computing the size of operation computer memory, and give definitions of a number of new concepts in engineering related to the application of methods of combinatorial analysis to the investigation of the performance of ACS. We propose a new method for estimating the efficiency of algorithms, characterized by extremal bounds.

The authors express their gratitude to all experts who helped in obtaining the results exposed in this book. In particular, O.V. Viskov, R.L. Graham, Ya. Demetrovich, Gy. Katona, Yu.V. Matiyasevich, S.G. Sal'nikov, and P. Erdös. The authors also thank A.F. Sidorenko, for supplementing the material of Chapter 3 by results concerning forbidden subgraphs and Ramsey numbers, and for taking part in writing the first two sections of Chapter 4. The authors express deep gratitude to A.A. Gushchin, V.K. Krivoshchekov, and A.A. Tsypkin for their great help in writing computer programs for obtaining the numerical results in Chapter 2.

Special thank of the authors goes to the reviewers, whose remarks not only helped in improving the book, but also influenced its structure.

Analysis Correspondence

(A Historical Perspective on Combinatorics)

The origins of systematic combinatorial research can be found in the works of Blaise Pascal and Pierre de Fermat. Questions about Chevalier de Méré's game led to the distinction between individual combinations and the calculation of favorable outcomes. Three chapters of Jakob Bernoulli's book '*Ars Conjectandi*' are devoted to the first systematic exposition of combinatorial facts. The works of Bernoulli and Gottfried Leibniz helped to establish combinatorics as an independent subject. In particular, Leibniz was first to attempt a full understanding of combinatorics in his dissertation '*Ars Combinatoria*', in which, apparently, the term 'combinatorics' originated.

In the Russian mathematics the term '`kombinatorika`' (combinatorics) was not immediately adopted, the preferred phrase being '`teoriya soedineniĭ`' (theory of joinings/combinations), which completely reflects the essence of the subject.

The basic objective notion in combinatorics is that of a correspondence. Combinatorics is the analysis of correspondences among properties of objects so as to study these properties and/or objects. Such analysis is difficult on account of the interdependence of such properties. Combinatorics consists in the study of correspondences and combinations of elementary mathematical objects—numbers, sets, and figures.

Methodologically, combinatorics is founded upon the three attributive properties of sets: distinctness, orderliness, and wholeness. Combining gives rise to all of the elementary combinatorial tools: distinctness—multiset, orderliness—permutation, and wholeness—partition. Objects of combinatorial study may include not only mathematical objects, but also any practical object, whether objects, people, acquaintanceship, or assertions. It is this very freedom in the selection of objects for investigation which provides the simplicity, accessibility, and practical significance of combinatorial results, and at times also their mystical breadth.

In the second half of the 19th century, the fundamentals of the theory of combinations began to be included in compulsory courses on algebra at high schools (`gimnaziya`) and technical schools in Russia and other countries. Deeper study of arrangements and combinations of objects was conducted in those branches of mathematics to which these objects belong: analysis, algebra, geometry, number theory, set theory, and logic. This, in turn, was reflected in the specificity and diversity of applications, and also in the formulation of the basic types of problems. Furthermore, all combinatorial topics are strongly interrelated, and unite in a single subject—combinatorics. General combinatorics encompasses all of these topics.

Towards the beginning of the 20th century, combinatorial analysis, as the mathematical analysis of functions of a discrete argument, 'occupied the ground between algebra and higher arithmetic', in the words of MacMahon, who also noted a tendency towards 'combinatorial attack in other territories'. The more pronounced this process, the more effective the methods of combinatorics became as they acquired new methods in the course of this very expansion. Two factors influence combinatorial investigations and their development in separate directions and on different topics: the type of subject, i.e. the selected object of investigation; and the type of problem, i.e. the selected aim of investigation. The choice depends on the needs observed and the possibilities available. The development of subject, in turn, enriches both the needs and the possibilities.

Elementary quantitative analysis of arrangements and combinations forms the basis of the traditional type of problem in combinatorics—enumeration. The development of enumeration is the main source of combinatorial analysis. The method of generating functions is commonly used in combinatorial analysis, and was historically the first method used. Developed by Leonard Euler for the theory of number partitions, this analytical method also turned out to be powerful in combinatorics. It has been developed into refined forms such as the method of Dirichlet generating functions, the method of trigonometric sums, and the method of characteristic functions: methods applicable not only in combinatorics and number theory. The method of generating functions was developed mainly for use in partition problems. One of the most brilliant moments in this development was the creation of the 'circle method', originally invented for counting the partitions of a given number.

Another type of problem in combinatorics is structural in nature. Such are best known in graph theory. Graph theory is a branch of combinatorics dealing with another kind of elementary relation on sets and systems of sets. However, the emergence of this branch occurred at a time when correspondence and relation had not yet been distinguished as independent mathematical concepts, but appeared only in terms of other notions, mainly of a geometrical and topological nature.

> 'I am not content with algebra, in that it yields neither the shortest proofs nor the most beautiful constructions of geometry. Consequently, in view of this I consider that we need yet another kind of analysis, geometric or linear, which deals directly with position, as algebra deals magnitude... Analysis situs. I think that I have at my disposal such a means, and that figures and even machines and motions can be represented using symbols, as algebra represents numbers and magnitudes.... I have yet to add a remark on the fact that I regard it possible to extend a characteristics to things inaccessible to perceptible imagination; but this is too important and too far fetched for me to be able to explain in this account in few words'.

So wrote Leibniz to Christiaan Huygens on September 8, 1679. In this letter, using the example of a kind of geometrical approach, Leibniz is searching for a general means for a formal operation with correspondences. The term 'situs' (position, location) itself may be understood as a correspondence between object and place. For the rest of his life, Leibniz did not abandon this dream. 15 years later he wrote to l'Hopital:

> '... I would like to have the possibility of realizing it, but dry and abstract thoughts excite me too much.... I was often not healthy in the past year, so that for a long time I had to restrain myself, although I did not succeed in this to the extent I would have wished'.

Leibniz's dream was ahead of its time, but, as turned out, not far ahead An amusing topological brain-teaser, the walking tour of the seven bridges of Königsberg, would be decisive. Euler derived necessary and sufficient conditions for the existence of such a walking tour in full generality, thereby lying the rudiments of graph theory. The original problem was as follows: Is it possible to walk across each bridge once and only once and return to the starting point? By representing the connected land regions by points and the bridges by lines, a graph can be drawn, and the question is that of the possibility of traversing the graph from point to point along the lines such that each edge is traversed once. On August 26, 1735, Euler presented a paper to the Academy of Sciences in St. Petersburg, 'Solutio problematis ad geometriam situs pertinentis'. In it he established local conditions for the realizability of such a traversal, which is now called an Euler tour. A graph has an Euler tour if and only if it is connected and has an even number of edges at each vertex. The graph of the Königsberg bridges does not satisfy this property. In the same paper Euler established that in any graph the sum of the degrees of the vertices equals twice the number of edges. Euler began his paper by referring to Leibniz as follows:

> 'In addition to that branch of geometry which is concerned with magnitudes and which has always received the greatest attention, there is another branch, previously almost unknown, which Leibniz first mentioned, calling it the geometry of positions (Geometria Situs). This branch is concerned only with the determination of position and its properties; it does not involve measurements, or calculations made with them ...'.

In this connection special significance attaches to an epistolary testimony of Euler (recently noted by A.E. Malykh and K.A. Rybnikov). On March 13, 1736, Euler wrote to J. Marioni about the problem of the bridges:

> 'This problem, although commonplace, seems to me, however, worthy of attention, since for its solution neither geometry, nor algebra, nor the combinatorial art suffices. Therefore a thought arose within me, which is not only accidentally related to the geometry of position (Geometria Situs) which was investigated by Leibniz in his time'.

This direct testimony indisputably stresses the inclusion within modern combinatorics not only of 'Ars Combinatoria', but also of 'Analysis Situs'. Thus, the notion of a graph as a system of two-element subsets (edges) of a set (of vertices) arose and was studied on the basis of its topological nature. The graph planarity criterion derived by Kazimierz Kuratowski extended the conception of graphs: A graph can be drawn in the plane by points and non-intersecting line segments if and only if it does not contain a subgraph homeomorphic to the graphs K_5 or $K_{3,3}$. This means that the 'topological nature' of a graph is completely determined by its set-theoretical structure. Therefore topological problems in graph theory from a separate subject matter, in particular, that of map coloring problems and problems on embedding graphs in manifolds. The graphical representation of arrangements and combinations by geometrical figures in combination with euclidean geometry led to the creation of the theory of matroids, and of combinatorial and of finite geometries.

The highly abstract nature of algebra, logic, and set theory not only brought about their application to relations amongst objects of arbitrary kind, but also made possible the solution of problems having to do with the realization of concrete structural phenomena, thereby lying the foundation of yet another direction: the algorithmic.

The characterization of the limiting possibilities of combinatorial relations forms the essence of yet another type, the extremal combinatorial problem, i.e., in general, the search for an answer to a question which can be stated in the words of Chebyshev:

> 'The major part of practical questions leads to problems involving largest or smallest magnitudes, which are totally new for science and only by solving these problems can satisfy the requirements of practice which always aim at the best, the most appropriate.... The coming together of theory and practice gives most favorable results, and not only practice gains by this; all of science develops under its influence, it discloses new subjects for investigation or new angles for considering long known subjects. Despite the high degree of development taken by science, practice clearly discloses the incompleteness of science in many respects; it poses problems which are essentially new for science, and in this manner calls for the search of totally new methods. If a theory can gain from new applications of old methods, or from new developments of it, then it gains even more by disclosing new methods, and in this case practice is a loyal guide of science.'

One of the first independent appearances of an extremal problem was geometrical, and dates back to 1611, when Johannes Kepler first described a means by which a sphere can e surrounded by twelve balls of the same radius such that all these balls touch the central sphere. 83 years later I. Newton and David Gregory quarreled about the question of how many balls of equal size one can distribute in this manner around a central sphere of the same radius. Newton asserted 12, and Gregory 13. The solution of this problem (twelve is correct) has taken over 200 years, and the search for a simple proof continues today. Studying the corpescular model of the structure of matter, Mikhail Lomonosov gave estimates of the coefficients of contraction of matter based on a density comparison for various manners of filling space with unit balls.

At approximately the same time, at the request of the Russian government, Christian Goldbach successfully attempted a mathematical approach to code-breaking, for which he was complemented by Counsellor Bestuzhev: 'There is a key for everything written in code, thanks to the art of Mr. Goldbach' (does this not underline the term 'key to a code'?).

Many problems of present-day coding theory can be formulated as extremal geometrical problems in Hamming space, e.g., the maximal cardinality of a uniform code of weight k and distance a is equal to the maximum number of vectors of norm k in a Hamming space for which the difference of any pair has norm at least a; clearly, this is an analog of the contact number. Thus, even in the period of development of the topic of extremal geometrical problems, the sphere of its possible uses already began to take shape. One of the earliest extremal number-theoretical results is due to James Sylvester, whose theorem asserts: Let r and s be mutually prime natural numbers, and let $n(r, s)$ be the largest integer n that cannot be represented in the form $n = ar + bs$, where a and b are non-negative integers. Then $n(r, s) = rs - r - s$. The same question for three or more mutually prime numbers is still open, and is called the *Frobenius problem*.

Extending the domain of application of theoretical combinatorial results leads to the emergence of another important type of problem, combinatorial modeling. Here, the choice of the most suitable combinatorial treatment of applied problems is determined by finite collection of their solutions. The high degree of abstraction of each combinatorial model makes it possible to investigate by this model a certain well-defined number of processes or phenomena in various domains of knowledge. Hence, the unification of such models into complexes, followed by the determination of a way of finding correspondence rules between the models, which, in turn, will depend on the problems to be solved by this complex of models, substantially extends the domain of their applicability.

In turn, this leads to another type of problem: the study of correspondences between distinct models. The main aim of this type is the creation of unified complexes of combinatorial models suitable for the adequate description not only of specialized practical problems, but also of processes and phenomena belonging to particular domains of knowledge.

We have to add a few comments about a relatively young topic, namely, algebraic combinatorics. This is first of all the study of compound structures (partial orders, binary relations and correspondences) by means of which one constructs in a suitable manner pure algebraic objects (incidence algebras, etc.). At first glance such algebras (introduced for arbitrary binary relations) may seem the ultimate in abstract constructions. It turns out that they are far from abstract, and Richard Stanley's result serves as a first substantiation of this. Nowadays there is a whole field arising from Stanley's theorem, which asserts that locally finite posets are isomorphic if and only if their incidence algebras are isomorphic (as abstract algebras). Thus, the inner structure of the original binary relation is uniquely determined by the corresponding algebra of functions. And this result holds for a large class of binary relations and algebras. Thus, the apparatus of incidence algebras may serve as a model for actual phenomena that can be adequately described by sets with binary relations. A concrete example is that of a combinatorial medium which includes such sets in the form of electronic networks and strategic analysis. Precisely the construction of algebras of functions over arbitrary binary relations has crucially extended the sphere of modeling actual phenomena, which previously had been restricted by the axioms of partial order. In particular, knowledge of the Möbius function of the incidence algebra over the binary relation provides full knowledge of the inner structure of this binary relation. The complexity of this structure corresponds to the complexity of this function and its forms, including forms which are nonstandard from the point of view of classical analysis. This young branch is so powerful that it merits more than an equal place among the general topics listed above, as it also changes the inner structure of combinatorics as a whole.

Combinatorics can serve as practice and as theory. In its formative period it proved practical for probability theory, confirming and suggesting its methods and laws; as theory it was outstandingly successful in solving problems. This remarkable duality also appears in extremal problems, which are not only a working tool for solving purely practical problems, but also characterize the efficiency of this solution, thus embodying a suitable standard of a basic criterion for truth: practice. The authors of this book concur with the thought expressed by Silvester: 'Number, place, and combination are three mutually interbreeding, but excellent spheres of thought to which all mathematical ideas can be reduced'. Finally, therefore, we conclude that in combinatorics the notion of correspondence is as fundamental as that of magnitude in algebra, number in number theory, or figure in geometry. Thus, alongside algebra, number theory, and geometry, combinatorics will ultimately occupy one of the 'atomic' places in the structural unity of all of mathematics.

The major part of this text is also our reaction to the paper of Robin J. Wilson: 'Analysis Situs' (*Graph Theory with Applications to Algorithms and Computer Science*, Wiley, New York, 1985, pp. 789–800) and the articles on combinatorics by V.N. Sachkov in the **Matematicheskaya Entsiklopediya** (Nauka, Moscow, 1985) [English translation (revised and annotated: *Encyclopaedia of Mathematics*, Kluwer, Dordrecht, 1988]. In particular, we have had to reiterate some essential statements from these works and are thus grateful to the authors of these publications for the opportunity to continue the discourse on this theme. Also, we express much gratitude to everyone who helped us write the final text in English: Rob Hoksbergen (the translator of the book), Kit Dodson, Bill Kocay, Luise Guy, and Jack McIntosh.

CHAPTER 1

Some information from combinatorics

This chapter gives definitions of the combinatorial notions necessary in the book. For a deeper study of these, we refer to the specialized handbooks [15], [53], [55]–[57], [60]–[63], [94], [100], [101].

1. Sets and operations with sets

1.1. The notions of a set and a multiset. A set is a whole, consisting of different parts. Of course, this description in words can hardly be regarded as a strict definition. The matter is that set is a categorial notion, not lending itself to a strict definition; its very nature allows various kinds of descriptions. The aim of these descriptions is to reflect the most important (attributive) properties of a set, in particular, distinctness of all parts of a set, nonorderability of the parts of a set, and the wholeness of a set.

We distinguish two types of parts of a set: elements and subsets. An element is taken to be an indivisible and nonempty part of a set, all other parts are taken to be subsets. Each element of a set can be regarded as a one-element subset. We distinguish a particular part, called the empty set (i.e. it does not contain any element) and denote it by $\emptyset$. We assume that every set has such a part.

Dropping the distinctness of elements of a set leads to the notion of multiset, i.e. a collection of elements some of which may be identical (indistinguishable). Each multiset can be represented by its basis, i.e. the set of all its different elements, and multiplicities, the numbers of repetition of each element of the basis of this multiset.

A handful of coins can be regarded as a set and as a multiset: If it contains coins with the same value, then these are indistinguishable as regards expenditure, i.e. they give a multiset, whereas at the same time, a numismatist may be interested in the dates on the coins, and if the dates are different on coins of the same value, then the handful of coins forms a set.

1.2. Notations. If a is an *element* of a set A, we say that a belongs to A and write $a \in A$; in the opposite case we write $a \notin A$. If B is a *subset* of A we write $B \subseteq A$. The relation $\subseteq$ of inclusion of subsets is reflexive ($A \subseteq A$) and transitive (if $B \subseteq A$ and $A \subseteq C$, then $B \subseteq C$). If $A \subseteq B$ and $B \subseteq A$, then $A = B$. A subset B is said to be a *proper subset* of A if $B \subseteq A$ and $A \neq B$. In such a case we write $B \subset A$.

The simplest numerical characteristic of a set as a whole is the amount of elements in it, i.e. the *cardinality of a set*. A set A is said to be *finite* if its cardinality is a nonnegative integer, denoted by $|A|$. If the number of elements of a set is not bounded, the set is said to be *infinite*. Let $|A| = n$ and $|B| = m$. If $B \subseteq A$, then $m \leq n$; moreover, if $B \subset A$, then $m < n$.

We can identify a set by its list of elements, $A = \{a_1, a_2, \dots\}$, where the order of the a_i's is immaterial. However, such an explicit specification of a set cannot always be given, or doing so may be inconvenient, e.g., the set $\mathbf{N}$ of natural numbers does not have an explicit listing, since it is infinite. In such cases, a set may be given by a description of the properties which uniquely determine whether elements do or do not belong to the set. This manner of specifying a set A can be correspondingly written as $A = \{a\colon a \text{ has property } R\}$, meaning that A consists of only those a that have the property $R(a) = R$, e.g., if $R(a)$ consists of the fact that a is a prime number (i.e. a cannot be represented as the product of two terms different from it and 1), then A is the set of all prime numbers. We can also specify a set recursively, when the next element is described in terms of previous ones. E.g., such a description of the set of natural numbers is

$$\mathbf{N} = \{i\colon \text{if the integer } i \in \mathbf{N}, \text{ then } i + 1 \in \mathbf{N}, i \geq 1, 1 \in \mathbf{N}\},$$

and such a description for the set of *Fibonacci numbers* is

$$F = \{p_i\colon p_i = p_{i-1} + p_{i-2}, p_1 = p_2 = 1, i = 3, 4, \dots\}.$$

The means for specifying a multiset are similar to those for specifying a set. E.g., the multiset $A = \{a, a, b, b, b, c\}$ has base $\{a, b, c\}$ and multiplicities $k(a) = 2$, $k(b) = 3$, $k(c) = 1$. The multiplicities of elements of the base of a multiset are sometimes written as superscripts; specifying A then corresponds to writing $A = \{a^2, b^3, c^1\}$. The list of multiplicities of a multiset $A = \{a^v, b^w, \dots\}$ is called its *first specification* and is denoted by $[A] = [v, w, \dots]$. According to this definition, the first specification can also be a multiset, and consists of natural numbers. The *second specification* of the multiset $A = \{a^v, b^w, \dots\}$ is the first specification of its first specification, i.e. $[[A]] = [[v, w, \dots]]$. This implies that if A is a set of m elements, then $[A] = [1^m]$, $[[A]] = [[1^m]] = \{m\}$.

In conclusion we note that every specification of a set must be correct. Nonobservance of this may lead to difficulties of the type of the Russell paradox. This paradox is usually illustrated by the example of the barber, defining the set of people he shaves as the set of all inhabitants of his village who do not shave themselves. In this specification of a set, it remains unclear whether or not the barber himself belongs to the set. Hence, any means of specifying a set must ensure its wholeness, whether given by elements, subsets, using operations, etc.

1.3. Operations with sets. The *intersection of two sets* X and Y is the set $X \cap Y$ of all elements that belong to both X and Y, i.e. $X \cap Y = \{x\colon x \in X \text{ and } x \in Y\}$. E.g., for $X = \{1, 2, 3\}$, $Y = \{2, 3, 4\}$, $X \cap Y = \{2, 3\}$, while for $A = \{1, 2\}$, $B = \{3, 4\}$, $A \cap B = \emptyset$ (such sets are said to be *disjoint*). Clearly, $X \cap \emptyset = \emptyset$. The intersection of two or more sets is commutative: $X \cap Y = Y \cap X$, and associative:

$$(X \cap Y) \cap Z = X \cap (Y \cap Z) = X \cap Y \cap Z.$$

The *union of two sets* X and Y is the set $X \cup Y$ of all elements that belong to either X or Y, i.e. $X \cup Y = \{x\colon x \in X \text{ or } x \in Y\}$. E.g., if $X = \{1,2,3\}$, $Y = \{2,3,4\}$, then $X \cup Y = \{1,2,3,4\}$; clearly, $X \cup \emptyset = X$. The union of two or more sets is commutative: $X \cup Y = Y \cup X$, and associative:

$$(X \cup Y) \cup Z = X \cup (Y \cup Z) = X \cup Y \cup Z.$$

The operations of union and intersection have the important property of *distributivity*:

$$(X \cap Y) \cup Z = (X \cap Y) \cup (X \cap Z), \qquad (X \cup Y) \cap Z = (X \cup Y) \cap (X \cup Z).$$

The *difference of two sets* X and Y is the set $X \setminus Y$ of all elements of X that do not belong to Y, i.e. $X \setminus Y = \{x\colon x \in X \text{ and } x \notin Y\}$. E.g., if $X = \{1,2,3\}$, $Y = \{2,3,4\}$, then $X \setminus Y = \{1\}$; clearly, $X \setminus \emptyset = X$ and $\emptyset \setminus X = \emptyset$. The definition of difference implies that $(X \setminus Y) \cup (X \cap Y) = X$.

The *symmetric difference of two sets* X and Y is the set $X \bigtriangleup Y$ of all elements of X that do not belong to Y and of all elements of Y that do not belong to X, i.e. $X \bigtriangleup Y = \{x\colon x \in X \text{ and } x \notin Y \text{ or } x \in Y \text{ and } x \notin X\}$. E.g., if $X = \{1,2,3\}$, $Y = \{2,3,4\}$, then $X \bigtriangleup Y = \{1,4\}$; clearly, $\emptyset \bigtriangleup X = X \bigtriangleup \emptyset = X$. The definition of symmetric difference implies that $X \bigtriangleup Y = X \cup Y \setminus X \cap Y$.

The *complement of a set* Y with respect to (in) a set X is defined only if $Y \subseteq X$, and then it is the set $\bar{Y} = X \setminus Y$. E.g., if $Y = \{2,3\}$, $X = \{1,2,3\}$, then the complement of Y in X is the set

$$\bar{Y} = X \setminus Y = \{1\}.$$

The *de Morgan laws*: if X and Y are subsets of a set Z, then $\overline{(X \cap Y)} = \bar{X} \cup \bar{Y}$, $\overline{(X \cup Y)} = \bar{X} \cap \bar{Y}$.

Sets $X_1, X_2, \ldots$ form a *cover of a set* X if $X \subseteq \cup_i X_i$; in this case the X_i are called *blocks* of the cover. E.g., a cover of the set of natural numbers is given by $\{1,2,\ldots\} \subset \cup_{i \geq 1}\{0, i, i+1\}$.

A *partition of a set* X is a representation by disjoint subsets of it: $X = X_1 \cup X_2 \cup \ldots$; $X_i \cap X_j = \emptyset$ $(i \neq j)$. E.g., $\{1,2,\ldots\} = \cup_{i=1}^{\infty}\{i\}$. The sets X_i are called *blocks* or *parts of the partition*. If the number of blocks is finite, it is called the *rank of the partition*. We can conveniently describe a partition by the list of its blocks, since, by definition, this list uniquely determines the partition; therefore the list is also called a partition. E.g., for $X = \{a, b, c\}$ the notation (a, bc) denotes the partition of X into two parts, a and bc, separated from each other by a comma.

The *specification* or *type of a partition* $X = X_1 \cup X_2 \cup \cdots \cup X_r$ is the list of cardinalities of its blocks, $[|X_1|, |X_2|, \ldots, |X_r|]$. E.g., the partition (a, bc) has type $[1, 2]$. A *subpartition* (or *splitting*) of a partition is a partition obtained by a partition of the blocks of the initial partition. E.g., the partition (a, b, c) is a subpartition of (a, bc). In other words, by taking unions of blocks in a subpartition we can always 'glue together' the initial partition. Finally, we distinguish between *ordered* and *nonordered partitions*, depending on whether or not the sequence of blocks is taken into account, where, moreover, all possible specifications distinct from the usual (nonordered) partition are agreed to be special.

The *sum rule* follows from the definition of partition of a set: for each partition of a finite set $X = X_1 \cup \cdots \cup X_r$, $X_i \cap X_j = \emptyset$ $(i \neq j)$, the following equality holds:

$$|X| = |X_1| + \cdots + |X_r|.$$

The *generalized sum rule* holds for a cover of a finite set $X \subseteq X_1 \cup \cdots \cup X_r$, and has the form:

$$|X| \leq |X_1| + \cdots + |X_r|.$$

The *product of sets* $X_1, \ldots, X_r$ is the set $\prod_{i=1}^{r} X_i = X_1 \cdot X_2 \cdot \ldots \cdot X_r$ consisting of all ordered tuples $(x_1, x_2, \ldots, x_r)$ where $x_i \in X_i$ $(i = 1, 2, \ldots, r)$. This product of sets is also called *direct* or *Cartesian*. If $X = \{1, 2\}$ and $Y = \{2, 3\}$, then $X \cdot Y = \{(1, 2), (1, 3), (2, 2), (2, 3)\}$. Hence, each element $(x_1, \ldots, x_r) \in \prod_{i=1}^{r} X_i$ of the direct product can be regarded as an r-dimensional vector, where $x_i \in X_i$ is the ith coordinate of this vector $(i = 1, 2, \ldots, r)$. It is convenient to set $X \cdot \emptyset = \emptyset$. The Cartesian product of n factors $X, \ldots, X$ is called the nth *Cartesian power* of X and is denoted by $X^{(n)}$. E.g., if $X = \{1, 2\}$, then

$$X^{(3)} = \{(1, 1, 1), (1, 1, 2), (1, 2, 1), (2, 1, 1), (1, 2, 2), (2, 1, 2), (2, 2, 1), (2, 2, 2)\}.$$

The *product rule* (which plays an important role in enumeration combinatorial problems): for arbitrary finite sets $X_1, X_2, \ldots, X_n$ the following equality holds:

$$|X_1 \cdot X_2 \cdot \ldots \cdot X_n| = |X_1| \cdot |X_2| \cdot \ldots \cdot |X_n|.$$

The *Boolean* is the set of all subsets of X, including the empty set $\emptyset$ and the set X itself. Thus, the elements of the Boolean, regarded as a set, are the subsets of X. E.g., the Boolean of $X = \{1, 2, 3\}$ consists of the sets $\{\emptyset\}$, $\{1\}$, $\{2\}$, $\{3\}$, $\{1, 2\}$, $\{1, 3\}$, $\{2, 3\}$, $\{1, 2, 3\}$. We denote the Boolean by 2^X or $\mathcal{P}(X)$; the notation 2^X is used in connection with the fact that for X finite, $|2^X| = 2^{|X|}$. In the Boolean, we can naturally distinguish the subsets (consisting of subsets of X) of equal cardinality: $C^k(X) = \{S \subseteq X : |S| = k\}$. Clearly, with this notation $\mathcal{P}(X) = \cup_{k=0}^{|X|} C^k(X)$. The sets $C^k(X)$ have cardinalities equal to the values of a binomial coefficient: if $|X| = n$, then

$$|C^k(X)| = C_n^k = \binom{n}{k} = \frac{n!}{k! \, (n-k)!}.$$

A *graph* on a set of vertices $S_n = \{a_1, \ldots, a_n\}$ is any subset G of the set $C^2(S_n)$, so that the elements of a graph $G \subseteq C^2(S_n)$ are the two-element subsets of S_n; these are called the *edges of the graph* G. Thus, each graph on a set of vertices $S_n = \{a_1, \ldots, a_n\}$ can be represented by the list of its edges, $G = \{(a_i, a_j), (a_k, a_l), \ldots\}$, where $(a_i, a_j) \in G$ if and only if the vertices a_i and a_j are joined by an edge of G. Consequently, every pair (a_i, a_j) in such a list can be interpreted as an edge.

The *complete graph* is the graph $K_n = C^2(S_n)$, so that $|K_n| = \binom{n}{2}$.

A *cycle* is a graph of the form $G = \{(a_1, a_2), (a_2, a_3), \ldots, (a_{k-1}, a_k), (a_k, a_1)\}$. A cycle is usually denoted by C_k; clearly $|C_k| = k$.

A *path* is a graph of the form $G = \{(a_1, a_2), (a_2, a_3), \ldots, (a_{k-1}, a_k)\}$. A path is usually denoted by P_k; clearly $|P_k| = k - 1$.

Usually, graphs are graphically depicted: the vertices of S_n are points, and the edges are lines joining the pairs of vertices forming an edge of the graph. E.g., in Figure 1.1 we have depicted the graph $G = \{(a_1, a_2), (a_1, a_4), (a_2, a_3), (a_2, a_4), (a_3, a_5)\}$

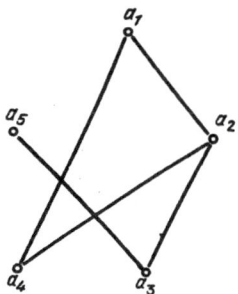

FIGURE 1.1

$(n = 5)$. This graph contains the complete subgraph K_3 (it is also the cycle C_3) on the three vertices a_1, a_2, a_4 and a path P_5, e.g., the path passing successively through a_5, a_3, a_2, a_1, a_4.

There are many different modifications of graphs.

The *directed graph*: the edges of G are ordered pairs of vertices.

The *multigraph*: the edges of G may repeat.

The *hypergraph*. A hypergraph on a set of vertices $S_n = \{a_1, \ldots, a_n\}$ is any subset G of $\mathcal{P}(S_n)$, so that the elements of a hypergraph $G \subseteq \mathcal{P}(S_n)$ are subsets of S_n, called *hyperedges* of G. Consequently, hyperedges of a hypergraph G may have cardinality larger than 2.

The *k-uniform hypergraph*, or *k-graph*: all edges of G have cardinality k.

The following are important numerical characteristics of graphs:

- The *vertex degree*: if $a \in S_n$, then $d_G(a) = |\{e \in G: a \in e\}|$, i.e. the vertex degree is the number of graph edges containing this vertex, in other words, are incident to this vertex.

- The *valency*: for vertex set S and a nonnegative integer q, the valency $v(S, q, G) = |\{e \in G: |S \cap e| = q\}|$ is the number of graph edges intersecting S in a fixed number q of vertices. Clearly, $v(a, 1, G) = d_G(a)$. Euler established that for each graph the degrees satisfy the identity $\sum_{i=1}^{n} d_G(a_i) = 2|G|$.

An *ordered partition* is a partition in which the sequence of blocks does matter. E.g., if $X = \{a, b, c\}$, then all ordered partitions of X are

with one block: (abc);
with two blocks: $(a, bc), (b, ac), (c, ab),$
: $(bc, a), (ac, b), (ab, c)$;
with three blocks: $(a, b, c), (a, c, b), (c, a, b),$
: $(b, a, c), (b, c, a), (c, b, a)$.

We denote the set of all ordered partitions of a set X by $T(X)$, and its cardinality by $T(|X|)$. We denote by $T^k(X)$ the set of ordered partitions consisting of k blocks, and by $T^k(|X|)$ the cardinality of this set. Now, if $|X| = n$, then

$$T(X) = \bigcup_{k=1}^{n} T^k(X), \qquad T(n) = \sum_{k=1}^{n} T^k(n).$$

As before, the notion of type is well-defined for ordered partitions, as the sequence of volumes of the blocks; therefore we denote by $T[n_1, \ldots, n_r]$ the set of ordered partitions of type $[n_1, \ldots, n_r]$, i.e. with blocks of volumes $n_1, \ldots, n_r$, respectively. E.g., the above described set of ordered partitions of $X = \{a, b, c\}$ with two blocks consists of the sets $T[1, 2]$ and $T[2, 1]$, having $3 \; (= T(1, 2) = T(2, 1))$ partitions in each of them. We denote the cardinality of the set $T[n_1, \ldots, n_r]$ by $T(n_1, \ldots, n_r)$. Now, if $|X| = n = \sum_{i=1}^{r} n_i$, then

$$T^r(X) = \cup_{(n_1, \ldots, n_r)} T[n_1, \ldots, n_r],$$
$$T^r(n) = \sum_{(n_1, \ldots, n_r)} T(n_1, \ldots, n_r).$$

Here, summation and union is over all types of partitions of rank r.

These numerical characteristics of ordered partitions may be calculated using the following formulas:

$$T^r(n_1, \ldots, n_r) = \binom{n!}{n_1, n_2, \ldots, n_r} = \frac{n!}{n_1! \, n_2! \ldots n_r!},$$

where $n!/(n_1! \, n_2! \ldots n_r!)$ is the multinomial coefficient;

$$T^r(n) = \sum_{\substack{n_1 + \cdots + n_r = n \\ n_i > 0}} \frac{n!}{n_1! \, n_2! \ldots n_r!} = \sum_{k=0}^{r} (-1)^{r-k} \binom{r}{k} k^n;$$

$$\sum_{\substack{n_1 + \cdots + n_r = n \\ n_i \geq 0}} \frac{n!}{n_1! \, n_2! \ldots n_r!} = r^n.$$

The *Bellean* is the set of all partitions of X. E.g., if $X = \{a, b, c\}$, then the Bellean of X consists of the partitions

    rank one: (abc);
    rank two: $(a, bc), (b, ac), (c, ab)$;
    rank three: (a, b, c).

Here the blocks are separated by commas. We assume that the partition blocks in a Bellean are not ordered, i.e. the partitions (c, ab) and (ab, c) are taken to be identical. We denote the Bellean by $B(X)$, its cardinality by $B(|X|)$, the set of all partitions with precisely k blocks by $B^k(X)$, and the cardinality of this set by $B^k(|X|)$. Thus, if $|X| = n$, then

$$B(X) = \bigcup_{k=1}^{n} B^k(X), \qquad B(n) = \sum_{k=1}^{n} B^k(n).$$

We denote the set of partitions of type $[n_1, \ldots, n_r]$ by $B[n_1, \ldots, n_r]$, and the number of partitions of a set X ($|X| = n$) of type $[n_1, \ldots, n_r]$, where $n = \sum_{i=1}^{r} n_i$, by $B(n_1, \ldots, n_r)$. Thus,

$$B^r(X) = \bigcup B(X_1, \ldots, X_r), \qquad B^r(n) = \sum B(n_1, \ldots, n_r),$$

where summation and union are over all possible types of partitions into r blocks.

We can calculate these numerical characteristics of the Bellean using the following formulas:

$$B(n_1, \ldots, n_r) = \frac{T(n_1, \ldots, n_r)}{r!},$$

$$B^r(n) = \frac{T^r(n)}{r!},$$

$B^k(n) = \sigma(n, k)$, where $\sigma(n, k)$ is the Stirling number of the second kind,

$$\sigma(n, k) = \sum_{j=0}^{k}(-1)^{k-j}\binom{k}{j}\frac{j^n}{n!}, \qquad k = 1, 2, \ldots, n,$$

$$\sigma(0, 0) = 1, \qquad \sigma(n, k) = 0 \quad (n < k),$$

$$x^n = \sum_{k=0}^{n}\sigma(n, k)x(x-1)\ldots(x-k+1),$$

$B(n)$ is Bell's number,

$$B(n) = \sum_{r=0}^{n}(r!)^{-1}\sum_{j=0}^{r}(-1)^{r-j}\binom{r}{j}j^n,$$

$$B(n+1) = \sum_{r=0}^{n}\binom{n}{r}B(n-r),$$

$$B(n) = \sum_{r=0}^{\infty}\frac{r^n}{r!e} \qquad \text{(Dobinski's formula)}.$$

1.4. Operations with multisets. For multisets we can introduce an addition operation which has no analog in classical set theory.

Addition of multisets. Suppose we are given multisets A and B: A has basis $S(A) = \{x, y, z, \ldots\}$ and multiplicities $[k_A(x), k_A(y), k_A(z), \ldots]$; B has basis $S(B) = \{x, y, z, \ldots\}$ and multiplicities $[k_B(x), k_B(y), k_B(z), \ldots]$. Then the *sum* $(A+B)$ of A and B is defined as the multiset with basis $S(A+B) = S(A) \cup S(B)$ and multiplicities

$$[k_{A+B}(x), k_{A+B}(y), k_{A+B}(z), \ldots] =$$
$$= [k_A(x) + k_B(x), k_A(y) + k_B(y), k_A(z) + k_B(z), \ldots],$$

i.e. under addition of multisets we take the union of their bases and add their multiplicities. E.g., if $A = \{a^2, b^3, c^1\}$, $B = \{a^1, c^5, d^4\}$, then $A+B = \{a^3, b^3, c^6, d^4\}$. Here, of course, elements absent in one basis but present in the other are interpreted to have multiplicity zero in this basis.

The definition of summation of multisets immediately implies a rule for computing the cardinality of their sum: if A and B are finite multisets, then

$$|A + B| = |A| + |B| = \sum_{a \in S(A)} k_A(a) + \sum_{b \in S(B)} k_B(b),$$

so that in the previous example $|A + B| = (2 + 3 + 1) + (1 + 5 + 4) = 16$.

Submultisets. We say that a multiset B with basis $S(B)$ is a submultiset of a multiset A with basis $S(A)$ if $S(B) \subseteq S(B)$ and for each $a \in S(B)$ we have $k_B(a) \leq k_A(a)$.

We denote inclusion of multisets by the same sign as for sets. E.g., if $A = \{a^2, b^7, c^1\}$, $B = \{a^1, b^5\}$, then $B \subset A$, since $S(B) = \{a, b\} \subset \{a, b, c\} = S(A)$ and $k_B(a) = 1 < 2 = k_A(a)$, $k_B(b) = 5 < 7 = k_A(b)$.

Operating with multisets. The operation of summation provides a very convenient technique for operating with sets and multisets, similar to the usual operation with numbers. Next to the summation of multisets, an important role is played in this technique by yet another notion, the wholeness operator which enables us to analytically operate with collections as with a wholeness.

Similarly as done for the Boolean, we take into consideration the set $C^k(A) = \{B : B \subseteq A, |B| = k\}$ of all k-element submultisets of a finite multiset A. E.g., if $A = \{a^2, b^3\}$ and $k = 3$, then $C^3(A)$ consists of three multisets: $\{a^2, b\}$, $\{a, b^2\}$, $\{b^3\}$.

The wholeness operator of a multiset A is the representation of A as a single element: $C(A) = C^{|A|}(A)$. E.g., if $A = \{a^2, b^3\}$, then $C(A) = (a, a, b, b, b)$. In case $A = S_n = \{a_1, \ldots, a_n\}$, i.e. A is a set, the wholeness operator $C(S_n) \neq S_n$, since according to the definition of $C(S_n)$ its cardinality $|C(S_n)| = \binom{n}{n} = 1$, while $|S_n| = n$. Thus, in essence the wholeness operator of a multiset A is the consideration of A as an integer (whole), and always $|C(A)| = 1$. From the above-said it is clear that if k is a nonnegative integer, then the notation $kC(a)$ must be understood as the k-times repetition of the element a.

For each multiset A the following equations hold (and are axiomatized):

$$A = \sum_{a \in A} C(a) = \sum_{a \in S(A)} k_A(a) C(a);$$

$$C(A) = \prod_{a \in A} C(a).$$

Here the product is to be understood as the usual product of sets. Precisely these equations allow us to formally operate with multisets. E.g., using them the above introduced operation of summation of multisets takes the following simple form:

$$A + B = \sum_{a \in S(A) \cup S(B)} (k_A(a) + k_B(a)) C(a).$$

This immediately implies the above given formula for the cardinality of a sum of multisets:

$$|A + B| = \left| \sum_{a \in S(A) \cup S(B)} (k_A(a) + k_B(a)) C(a) \right| =$$

$$= \sum_{a \in S(A) \cup S(B)} |(k_A(a) + k_B(a)) C(a)| =$$

$$= \sum_{a \in S(A) \cup S(B)} (k_A(a) + k_B(a)) |C(a)| =$$

$$= \sum_{a \in S(A) \cup S(B)} (k_A(a) + k_B(a)).$$

Moreover, the operations of intersection and union of multisets can be described in a very simple manner. For this we introduce the notation $\wedge = \min$, $\vee = \max$.

Then

$$A \cap B = \sum_{a \in S(A) \cap S(B)} (k_A(a) \wedge k_B(a))C(a),$$

$$A \cup B = \sum_{a \in S(A) \cup S(B)} (k_A(a) \vee k_B(a))C(a).$$

We will define the *product of multisets* in such a way that the following product rule holds:

if A and B are finite multisets, then $|AB| = |A| \cdot |B|$.

Starting from this requirement we assume that if A and B are multisets, then their product is the multiset

$$(A \cdot B) = \sum_{a \in A} \sum_{b \in B} C(C(a) + C(b)).$$

From this definition we obtain, as required:

$$|A \cdot B| = \left| \sum_{a \in A} \sum_{b \in B} C(C(a) + C(b)) \right| = \sum_{a \in A} \sum_{b \in B} |C(C(a) + C(b))| =$$

$$= \sum_{a \in A} \sum_{b \in B} 1 = \left(\sum_{a \in A} 1 \right) \left(\sum_{b \in B} 1 \right) = |A| \cdot |B|.$$

E.g., if $A = \{a^2, b^1\}$, $B = \{a^1, b^2\}$, then their product consists of nine pairs of elements: $A \cdot B = \{(a, a), (a, b), (a, b), (b, a), (b, b)\} = \{(a, a)^2, (a, b)^4, (b, a)^1, (b, b)^2\}$. Another example of a product of multisets is the ordinary product of natural numbers, since each natural number can be represented as the multiset consisting of the units, i.e. n has basis $\{1\}$ and multiplicity n.

The *Boolean of a multiset*. Let A be the finite multiset with basis $S(A) = \{a_1, a_2, \ldots, a_r\}$ and first specification $[k_1, k_2, \ldots, k_r]$, i.e. the element a_i occurs k_i times in A $(i = 1, 2, \ldots, r)$ and the cardinality of this multiset is $|A| = \sum_{i=1}^{r} k_i = n$, so that

$$A = \sum_{a \in A} C(a) = \sum_{a \in A} k_A(a)C(a) = \sum_{i=1}^{r} k_i C(a_i).$$

The Boolean of A is the set of its submultisets, including the empty set and the multiset A itself. We denote this Boolean by $\mathcal{P}(A)$. From the definition of $C^k(A)$ we have $\mathcal{P}(A) = \cup_{k=0}^{n} C^k(A) = \sum_{k=0}^{n} C^k(A)$, hence, using the technique of operating with multisets we obtain

$$\mathcal{P}(A) = \sum_{k=0}^{n} C^k(A) = \sum_{k=0}^{n} \sum_{\substack{j=0 \\ j_1 + \cdots + j_r = k}} \cdots \sum_{j=0}^{k_r} \prod_{i=1}^{r} C^j(k_i C(a_i)) =$$

$$= \sum_{j=0}^{k_1} \cdots \sum_{j=0}^{k_r} \prod_{i=1}^{r} C^j(k_i C(a_i)) = \prod_{i=1}^{r} \sum_{j=0}^{k_i} C^j(k_i C(a_i)) = \prod_{i=1}^{r} \mathcal{P}(k_i C(a_i)).$$

Thus, the Boolean of a multiset can be represented as the direct product of the Booleans $\mathcal{P}(k_i C(a_i))$ of multisets consisting of the single elements a_i, repeated k_i

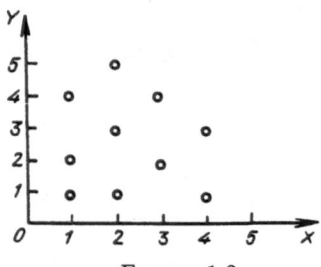

FIGURE 1.2

times. The Boolean of such a multiset obviously consists of the $k_i + 1$ submultisets $\{0\}, \{a_i\}, \{a_i, a_i\}, \ldots, \{a_i^{k_i}\}$, i.e. $|\mathcal{P}(k_i C(a_i))| = k_i + 1$. Consequently,

$$|\mathcal{P}(A)| = \left| \prod_{i=1}^{r} \mathcal{P}(k_i C(a_i)) \right| = \prod_{i=1}^{r} |\mathcal{P}(k_i C(a_i))| = \prod_{i=1}^{r} (k_i + 1).$$

For the set $A = S_n = \{a_1, \ldots, a_n\}$ this formula gives the formula $|\mathcal{P}(S_n)| = 2^n$, which is already known to us. Moreover, it is useful to note the case $A = kS_n = \{a_1^k, \ldots, a_n^k\}$, for which we have

$$|C^k(kS_n)| = \binom{n+k-1}{k}.$$

2. Correspondences between sets

2.1. A *correspondence* between sets X and Y is any subset $Z \subseteq X \times Y$ given in advance. If $(x, y) \in Z$, then we say that *the element y corresponds to the element x*, or that *the elements x and y are in correspondence Z*, and we write xZy or $Z(x, y)$. The element y is called the *image* of x, and x is called the *pre-image* (or *inverse image*) of y under the correspondence Z. If, on the other hand, $(x, y) \notin Z$, we write $x\bar{Z}y$. E.g., if $X = \{1, 2, 3\}$, $Y = \{3, 4, 5\}$ and the correspondence Z consists of the fact that $x + y$ is a prime number ($x \in X$, $y \in Y$), then $Z = \{(1, 4), (2, 3), (2, 5), (3, 4)\}$. Here the elements 1 and 3 of X have the same image (namely 4), while $2 \in X$ has two images (namely, 3 and 5); similarly, 3 and 5 in Y have the same pre-image (2 and 2), while $4 \in Y$ has two pre-images (1 and 3).

A subset $Z \subset \prod_{i=1}^{n} X_i$ is called an *n-place correspondence* between the sets X_i ($i = 1, 2, \ldots, n$). Thus, each vector $(x_1, x_2, \ldots, x_n)$ can be regarded as an element of a certain *n*-place correspondence. In particular, this shows that a correspondence can be geometrically given, by depicting the corresponding set of vectors in the Cartesian product of the sets. Other means for specifying correspondences are also useful: graphically and tabularly. We consider these three means in concrete examples. The correspondence $Z = \{x + y \text{ are prime numbers}\}$ is represented in Figure 1.2. Here a given point with coordinates (x, y) means that $(x, y) \in Z$. For the same X and Y, suppose the correspondence $Z \subseteq XY$ is given by the rule: $(x, y) \in Z$ if and only if $x + y$ is even. The geometrical (a), graphical (b), and tabular (c) specifications of Z are given in Figure 1.3. It is clear from Figure 1.3 that in the geometrical specification of Z the relation $(x, y) \in Z$ means that a point in the plane is drawn; in the graphical specification a line segment is drawn; and in the

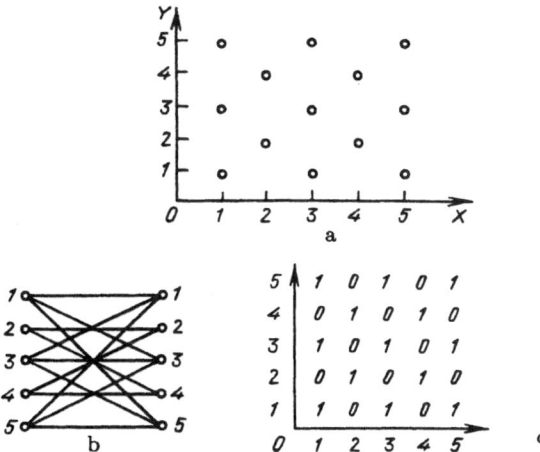

FIGURE 1.3. a b c

tabular specification a 1 is placed. This table is called the *incidence matrix* of the correspondence.

The *complete image* of an element $x \in X$ under a correspondence $Z \subseteq XY$ is the subset $\{y: y \in Y, xZy\} \subset Y$; similarly, the *complete pre-image* of an element $y \in Y$ under a correspondence $Z \subseteq XY$ is the subset $\{x: x \in X, xZy\} \subset X$. E.g., for the correspondence in Figure 1.2 it is clear that the complete pre-image of 5 consists of the single element 2, and the complete pre-image of 4 consists of the two elements 1 and 3. We can use correspondences to specify multisets. E.g., the multiset $A = \{a^2, b^1, c^4, d^3\}$ can be given by the correspondence $Z \subset \{a, b, c, d\}\{1, 2, 3, 4\}$ defined by the rule: $(x, n) \in Z \Leftrightarrow k_A(x) = n$, i.e. when A contains exactly n copies of x, where $x \in \{a, b, c, d\}$ and $n \in \{1, 2, 3, 4\}$.

2.2. A map ϕ from a set X into a set Y is a correspondence $Z \subset XY$ such that for each $x \in X$ there is at most one $y \in Y$ for which xZy. This correspondence between x and y under ϕ is written as the equality $y = \phi(x)$, and in this case the map has corresponding notation $\phi: X \to Y$. The set $X' \subset X$ of those $x \in X$ for which there is exactly one $y \in Y$ such that xZy is called the *domain of definition* of ϕ, and the set $Y' \subset Y$ of $y \in Y$ for which there are $x \in X$ such that xZy is called the *range of values* of ϕ. If $X_1 = X$, $Y_1 = Y$, and $\phi(x) = \phi_1(x)$ for all $x \in X$, then we say that the maps ϕ and ϕ_1 coincide, i.e. $\phi: X \to Y$ is equal to $\phi_1: X_1 \to Y_1$. E.g., if the sets $X = \{2, 3, 4\}$ and $Y = \{3, 4, 5\}$ are given, and the map $\phi: X \to Y$ assigns to an element of X a multiple of it in Y, then the correspondence Z corresponding to ϕ is $Z = \{(2, 4), (3, 3), (4, 4)\}$. A direct verification shows that Z is in fact a map, since each element in X is set to correspond to a unique element of Y, $(2, 3, 4)$ go to $(4, 3, 4)$, respectively. We have to make the following remark: if the rule defining ϕ is extended to a larger subset of the integers, then it is easily seen that Z need not always specify a map, since there may be elements in X having more than one image. In the example above, the domain of definition is the whole set X, and the range of values is $Y' = \{3, 4\}$.

The set $\phi^{-1}(y) = \{x: y = \phi(x), x \in X\}$ is called the *complete pre-image* of the element y under the map ϕ.

Let $X = \{x_1, x_2, \ldots, x_n\}$. Then $\phi\colon X \to Y$ can be represented as

$$\phi = \begin{pmatrix} x_1, & x_2, & \cdots, & x_n \\ \phi(x_1), & \phi(x_2), & \cdots, & \phi(x_n) \end{pmatrix},$$

where $\phi(x_i) \in Y$ $(i = 1, 2, \ldots, n)$. E.g., if $X = \{1, 2, 3, 4, 5\}$, $Y = \{a, b\}$, and

$$\phi = \begin{pmatrix} 1, & 2, & 3, & 4, & 5 \\ a, & a, & b, & b, & a \end{pmatrix},$$

then the complete pre-image of a is $\phi^{-1}(a) = \{1, 2, 5\}$, and the complete pre-image of b is $\phi^{-1}(b) = \{3, 4\}$.

The set $\phi(X) = \{\phi(x) \subset Y\colon x \in X\}$ is called the *complete image* of the domain of definition of the map ϕ.

If $\phi\colon X \to Y$ is such that $\phi(X) = Y$, then we say that ϕ maps X *onto* Y; in this case there is for each $y \in Y$ an $x \in X$ such that $y = \phi(x)$, and the condition $\phi^{-1}(y) \neq \emptyset$ holds. For finite X and Y the equality $\phi(X) = Y$ means that $|X| \geq |Y|$.

If $\phi\colon X \to Y$ is such that for the complete pre-image of each $y \in Y$ we have $|\phi^{-1}(y)| \leq 1$, then for finite X and Y we have $|X| \leq |Y|$. If we have $|\phi^{-1}(y)| = 1$ for all $y \in Y$, i.e. for each $y \in Y$ the condition $y = \phi(x)$ determines a unique element $x \in X$, then we say that ϕ sets up a *bijective correspondence* between X and Y. For finite sets X and Y we then have $|X| = |Y|$. In this case ϕ is called a *bijective map* or *bijection*. E.g., Let $X = \{1, 2, 3\}$, $Y = \{a, b, c\}$, and let ϕ be such that $\phi(1) = b$, $\phi(2) = c$, $\phi(3) = a$, then ϕ is a bijection between X and Y.

A *permutation* of a finite set is a bijection of this set onto itself. E.g., if $X = \{1, 2, 3\}$, then

$$\begin{pmatrix} 1 & 2 & 3 \\ 3 & 1 & 2 \end{pmatrix}$$

is a permutation. If $|X| = n$, then the number of all permutations of the Y is

$$n! = 1 \cdot 2 \cdot \ldots \cdot n.$$

In fact, the first element can be chosen from any of the n elements of X, the second from the remaining $n - 1$, the third from the remaining $n - 2$, etc. Permutations can be multiplied by the rule

$$\begin{pmatrix} 1 & 2 & \cdots & n \\ i_1 & i_2 & \cdots & i_n \end{pmatrix} \begin{pmatrix} 1 & 2 & \cdots & n \\ j_1 & j_2 & \cdots & j_n \end{pmatrix} = \begin{pmatrix} 1 & 2 & \cdots & n \\ j_{i_1} & j_{i_2} & \cdots & j_{i_n} \end{pmatrix}.$$

This multiplication corresponds to superposition of the maps corresponding to the permutations to be multiplied; in other words, to successive application of these two maps. This means that if the map

$$\phi_i = \begin{pmatrix} 1 & 2 & \cdots & n \\ i_1 & i_2 & \cdots & i_n \end{pmatrix}$$

puts the elements $(1, 2, \ldots, n)$ in the order $(i_1, i_2, \ldots, i_n)$, and the map

$$\phi_j = \begin{pmatrix} 1 & 2 & \cdots & n \\ j_1 & j_2 & \cdots & j_n \end{pmatrix}$$

puts the elements $(1, 2, \ldots, n)$ in the order $(j_1, j_2, \ldots, j_n)$, then the map $\phi_i \phi_j$ first puts the elements $(1, 2, \ldots, n)$ in the order $(i_1, i_2, \ldots, i_n)$, and then puts this order in the order $(j_{i_1}, j_{i_2}, \ldots, j_{i_n})$. E.g., if $X = \{1, 2, 3\}$ and

$$\phi_i = \begin{pmatrix} 1 & 2 & 3 \\ 3 & 1 & 2 \end{pmatrix}, \qquad \phi_j = \begin{pmatrix} 1 & 2 & 3 \\ 3 & 2 & 1 \end{pmatrix},$$

then

$$\phi_i \phi_j = \begin{pmatrix} 1 & 2 & 3 \\ 1 & 3 & 2 \end{pmatrix}.$$

There is a unit permutation, not displacing any element:

$$e = \begin{pmatrix} 1 & 2 & \ldots & n \\ 1 & 2 & \ldots & n \end{pmatrix}.$$

It is easily verified that for every permutation ϕ we have $e\phi = \phi e = \phi$. There are also permutations not leaving any element in place, e.g. the permutation

$$\begin{pmatrix} 1 & 2 & \ldots & n-1 & n \\ 2 & 3 & \ldots & n & 1 \end{pmatrix}$$

represents a shift, or is cyclic.

The *restriction* of a map $\phi \colon X \to X$ to a subset $Y \subseteq X$ is the map $\phi \colon Y \to X$, i.e. the same map ϕ, but on a smaller domain of definition. The restriction of ϕ to $Y \subseteq X$ is called a *cycle* if $\phi(Y) = Y$ and for some partition $Y = Y_1 \cup Y_2$, $Y_1 \cap Y_2 = \emptyset$, $Y_i \neq \emptyset$, $i = 1, 2$, there is an element $y \in Y_1$ such that $\phi(y) \in Y_2$; in this case Y_1 is called an *orbit of the cycle*.

A permutation ϕ acting on a set $X = \{x_1, x_2, \ldots\}$ is called a *transposition* if it leaves all elements, except two, in place, while these two elements change places. E.g., the permutation

$$\begin{pmatrix} 1 & 2 & 3 & 4 & 5 & 6 \\ 1 & 2 & 5 & 4 & 3 & 6 \end{pmatrix}$$

is a transposition on the set $\{1, 2, 3, 4, 5, 6\}$. Thus, a transposition has a cycle of cardinality two; in this case it is the cycle

$$\begin{pmatrix} 3 & 5 \\ 5 & 3 \end{pmatrix}.$$

If an orbit of a cycle consist of a single element, this element is said to be *fixed*. For cycles

$$\begin{pmatrix} 1 & 2 & \ldots & i_1 & \ldots & i_2 & \ldots & i_r & \ldots & n \\ 1 & 2 & \ldots & i_2 & \ldots & i_3 & \ldots & i_1 & \ldots & n \end{pmatrix}$$

we use the convenient row notation $(i_1 i_2 \ldots i_r)$, while for $n = 1$, $(i_1) = e$. E.g, if $X = \{1, 2, 3, 4, 5, 6\}$ and

$$\phi = \begin{pmatrix} 1 & 2 & 3 & 4 & 5 & 6 \\ 3 & 1 & 2 & 4 & 6 & 5 \end{pmatrix}$$

then the restriction of ϕ to $Y = \{1, 2, 3\}$ has the form

$$\begin{pmatrix} 1 & 2 & 3 \\ 3 & 1 & 2 \end{pmatrix}.$$

and is a cycle, the element 4 is fixed, and the cycle

$$\begin{pmatrix} 5 & 6 \\ 6 & 5 \end{pmatrix}$$

is a transposition.

Thus, each permutation can be represented as the product of its cycles, and generates a partition of X into orbits. In turn, each cycle can be represented as a product of transpositions, e.g.

$$\begin{pmatrix} 1 & 2 & 3 & 4 & 5 & 6 \\ 3 & 1 & 2 & 4 & 6 & 5 \end{pmatrix} = (1\,3\,2)(4)(5\,6) = (1\,3)(1\,2)(4)(5\,6).$$

A *rearrangement* of a finite set is the complete image of a bijection of this set onto itself. E.g., in the previous example, $(3\,1\,2\,4\,6\,5)$ is a rearrangement, i.e. the lower row of a permutation is a rearrangement if the order in the upper row is fixed.

Two graphs $G(S_n)$ and $G'(S_n)$ with vertex set S_n are said to be *isomorphic* if there is an enumeration of the vertices of one such that the lists of their edges coincide. E.g., the graphs $G = \{(a_1, a_2), (a_1, a_4)\}$ and $G' = \{(a_2, a_4), (a_2, a_3)\}$ are isomorphic, since the enumeration of the vertices of the first graph given by the permutation $(1\,2\,3)$ makes the first graph identical to the second. Since each enumeration of vertices is uniquely determined by a permutation, we also say that the graphs $G(S_n)$ and $G'(S_n)$ with vertex set S_n are isomorphic if there is a permutation π of S_n satisfying

$$G(S_n) = G'(\pi(S_n)).$$

A *sequence of objects* of whatever kind is a map from the set of natural numbers into the set of all these objects. E.g., $\{1, 3, 5, 7, 9, \dots\}$ is the sequence of all odd numbers, and the map consists of putting 1 at the first place, 3 at the second, 5 at the third, etc. Passing to the two-row notation, this map can be written as

$$\begin{matrix} 1, & 2, & 3, & 4, & 5, & \dots \\ 1, & 3, & 5, & 7, & 9, & \dots \end{matrix}$$

Thus, a 'sequence' supposes an ordered list of elements, representing a functional dependence between its elements and the natural numbers. In the example given this dependence is $\phi(n) = 2n - 1$. The notion of map allows us to introduce the notions of operation and function.

2.3. Operations. We say that an n-place *operation* λ is given on a set X if we are given a map $\lambda \colon X^{(n)} \to X$ which associates to a vector $(x_1, x_2, \dots, x_n) \in X^{(n)}$ a single, unique element $x \in X$. We write this as

$$x = \lambda(x_1, x_2, \dots, x_n).$$

Two-place, or binary, operations are the most widespread. A *binary operation* on a set X is a rule associating to an element of $X^{(2)}$ at most one element of X. Binary operations are usually given a special notation, and in the general case we will write $x = x_1 T x_2$. E.g., if the addition operation is given on the set $X = \{1, 2, 3\}$, then to only two pairs of elements an element of X is associated $(1 + 1 = 2, 1 + 2 = 3)$, since the sums of all other pairs do not belong to X.

We say that a set X is *closed* under a binary operation given on it if to each element of $X^{(2)}$ is associated an element of X. E.g., if $X = \mathbf{N} = \{1, 2, \dots\}$ is the set of natural numbers and $T = (+)$ is the addition operation, then the result of this operation is the sum $x = x_1 + x_2$, which obviously also belongs to $\mathbf{N} = X$. This means that the set of natural numbers is closed under the addition operation. We can similarly be convinced that it is closed under multiplication, and not closed under the operations of subtraction and division. In this way we can represent each binary operation on a set by a ternary, i.e. three-placed, correspondence on this set.

A binary operation T on a set X is said to be

associative: if for any $x, y \in X$ the following condition holds:
$$(xTy)Tz = xT(yTz);$$

commutative: if for any $x, y \in X$ the following condition holds:
$$xTy = yTx;$$

distributive with respect to an *operation* ∂: if for any $x, y, z \in X$ the following equalities hold:
$$xT(y\partial z) = (xTy)\partial(xTz), \qquad (y\partial z)Tx = (yTx)\partial(zTx).$$

An element e is said to be the *unit element*, or *neutral element*, under a binary operation T if for all $x \in X$, $eTx = xTe = x$. E.g., the addition and multiplication operations on the set of real numbers are associative and commutative. Multiplication is distributive with respect to addition. The unit elements under multiplication and addition are 1 and 0, respectively.

The set of all permutations of the elements in $\{1, 2, \dots, n\}$ is closed under the above introduced operation of multiplication of permutations; the unit element of this multiplication is
$$e = \begin{pmatrix} 1 & 2 & \dots & n \\ 1 & 2 & \dots & n \end{pmatrix}.$$
This operation is associative, but not commutative.

2.4. Functions. By *function* we mean a map into the real or complex numbers. We list the simplest combinatorial functions:

factorial: if n is a natural number, then $n! = n(n-1)\dots 1$; $0! = 1$; if n and m are natural numbers, then
$$(n)_m = \begin{cases} n(n-1)\dots(n-m+1), & m \le n, \\ 0, & m > n; \end{cases}$$

binomial coefficient: if n, k are integers, then
$$\binom{n}{k} = C_n^k = \frac{n!}{k!\,(n-k)!}, \qquad 0 \le k \le n;$$

multinomial coefficient: if $n = k_1 + k_2 + \dots + k_t$, where $k_1, k_2, \dots, k_t$ are integers, then
$$\binom{n}{k_1 \dots k_t} = \frac{n!}{k_1!\,k_2!\dots k_t!}, \qquad k_i \ge 0$$

(this is also called the *polynomial coefficient*);

entier and fractional part: if x is a real number, then we denote by $[x]$ its *entier*, i.e. the largest integer not exceeding x; e.g., $[5.3] = 5$, $[-5.3] = -6$. So, x is an integer if and only if $x = [x]$. We denote by $]x[$ the smallest integer not less than x; e.g., $]5.3[= 6$, $]-5.3[= -5$. We have $]x[= -[-x]$. The *fractional part* of x is the number $\{x\} = x - [x]$; e.g., $\{7\} = 0$, $\{2.6\} = 0.6$, $\{-4.75\} = 0.25$. Sometimes we will use another notation: $\lfloor x \rfloor = [x]$ ('floor') and $\lceil x \rceil =]x[$ ('ceiling');

indicator function:

$$\chi\{\text{assertion}\} = \begin{cases} 1 & \text{if the assertion is true,} \\ 0 & \text{if the assertion is false.} \end{cases}$$

2.5. A *relation* is a correspondence between identical sets. A two-place relation is also said to be binary. Examples of and means for specifying a binary relation are given in Figure 1.3.

We distinguish the following properties of a binary relation $R \subseteq X^{(2)}$ on a set X:

reflexivity: xRx for all $x \in X$;

anti-reflexivity: $x\bar{R}x$ for all $x \in X$;

symmetry: xRy implies yRx for all $x, y \in X$;

anti-symmetry: xRy and yRx imply $x = y$ for all $x, y \in X$;

transitivity: xRy and yRz imply xRz for all $x, y, z \in X$; 'un serviteur d'un serviteur de moi ne pas un serviteur de moi' is an example of an intransitive relation;

dichotomy: for any $x, y \in X$, either xRy or yRx.[1]

Relations emerge often in practice. E.g., acquaintance of people is reflexive and symmetric, but not always transitive. Every hierarchy is also a binary relation, so that various ordered sets may be conveniently characterized by binary relations.

2.6. An *ordered set* is a pair (X, R), where X is a set and R a binary relation, $R \subseteq X^{(2)}$. If we have xRy for $x, y \in X$, i.e. $(x, y) \in R$, then we conveniently interpret this as x 'being larger' than y in the sense of the relation R. If neither xRy nor yRx, then x and y are *incomparable elements* of (X, R). We consider the following basic types of ordered sets:

totally unordered: a pair (X, R) with $R = \emptyset$;

totally ordered: a pair (X, R) with R reflexive, anti-symmetric, transitive, and total. An example of such a set are the natural numbers, ordered by magnitude, i.e. by the relation $\leq$;

partially ordered: an ordered set (X, R) with R reflexive, anti-symmetric, and transitive.

In ordered sets we can conveniently distinguish specific individual elements and subsets; we give some of these.

largest in (X, R): an element $w \in X$ such that for all $x \in X$ we have wRx, i.e. w is 'larger' than every element of X. Largest elements are sometimes simply called units, and are denoted by 1.

----

[1] R is then also said to be *total* or *complete*.

FIGURE 1.4

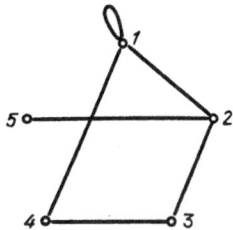

FIGURE 1.5

maximal in (X, R): an element $w \in X$ such that X does not contain an element x satisfying xRw, i.e. X does not contain an element 'larger' than w. In other words, every $x \in X$ is either incomparable with w or is 'smaller' than w.

least in (X, R): an element $v \in X$ such that for all $x \in X$ we have xRv, i.e. v is 'smaller' than every element of X. Least elements are sometimes simply called zeros, and are denoted by 0.

minimal in (X, R): an element $v \in X$ such that X does not contain an element x satisfying vRx, i.e. X does not contain an element 'smaller' than v. In other words, every $x \in X$ is either incomparable with v or is 'larger' than v.

We say that an element x *covers* an element y if xRy and there is no $z \in X$, distinct from x and y, such that xRz and zRy. *Atoms* are elements covering 0, and *co-atoms* are elements covered by 1. We say that two elements x and y are *incomparable* if $(x, y) \notin R$ and $(y, x) \notin R$, i.e. neither x is 'larger' than y, nor y is 'larger' than x.

For $x, y \in X$, the *interval* $[x, y]$ is the subset $[x, y] = \{z: z \in X, yRz, zRx\}$, i.e. this interval is the set of elements that are at the same time 'smaller' than y and 'larger' than x. An ordered set (X, R) is said to be *locally finite* if every interval in it is finite.

A *chain* in an ordered set (X, R) is a sequence of elements $a_1, a_2, \ldots, a_k, \ldots$ in X such that $a_1 R a_2, a_2 R a_3, \ldots, a_{k-1} R a_k, \ldots$.

The *length of a finite chain* is the number of its terms minus 1.

An *antichain* is a subset of an ordered set consisting of pairwise incomparable elements only.

An ordered set (X, R) can be conveniently depicted in graphical form: the set X gives the points, and the relation R the oriented lines, directed from x to y if xRy.

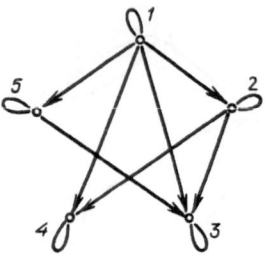

FIGURE 1.6

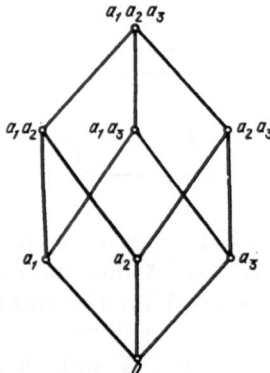

FIGURE 1.7

The graph thus constructed uniquely determines (X, R); e.g., the graph in Figure 1.4 determines a binary relation on $X^{(2)}$, where $X = \{1, 2, 3, 4, 5\}$. In a number of cases this graph can be simplified. E.g., if it is known that (X, R) is reflexive, then the arc from x to x may be omitted. Similarly, for a symmetric R the graph can be drawn undirected. E.g., the graph of the relation $Z = \{x + y \text{ are prime numbers}\} \subseteq X^{(2)}$, which has already been considered above, where $X = \{1, 2, 3, 4, 5\}$, is drawn in Figure 1.5.

For depicting an ordered set we use the *Hasse diagram*: a graph in which the edge (x, y) is present if and only if y covers x. Since R is transitive, in this graph the condition $(x, y) \in R$ gives a path from y to x. To orient the edges of the Hasse diagram it is convenient to draw minimal elements in its lower part, and maximal elements in its upper part. E.g., in Figure 1.6 we have drawn a partially ordered set with two minimal elements $(3, 4)$ and one largest (1). More precisely, this graph represents the following partial order: $R = \{(1, 1), (2, 2), (3, 3), (4, 4), (5, 5), (1, 2), (1, 3), (1, 4), (1, 5), (2, 3), (2, 4), (5, 3)\} \subseteq X^{(2)}$, where $X = \{1, 2, 3, 4, 5\}$. Hence the Hasse diagram is obtained by removing the edges $(1, 3)$ and $(1, 4)$ and all loops.

We give the basic combinatorial examples of ordered sets.

Boolean, ordered by inclusion of subsets: We consider the Boolean of $S_n = \{a_1, a_2, \ldots, a_n\}$, $\mathcal{P}(S_n) = \cup_{k=0}^{n} C^k(S_n)$, and introduce a binary relation

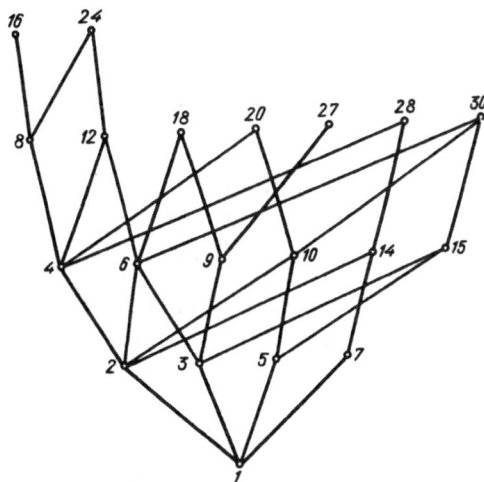

FIGURE 1.8

$\subseteq$ on it according to the rule: For $X, Y \in \mathcal{P}(S_n)$ we have $X \subseteq Y$ if and only if X is a subset of Y. This binary relation is a partial order relation.

There is a least element in the Boolean, i.e. a subset of all other elements of the Boolean. Clearly, such is the empty set $\emptyset$. There is also a largest element, i.e. containing all other elements of the Boolean. Clearly, this is the set S_n itself. Obviously, elements of the Boolean contained in a single $C^k(S_n)$ cannot be included into each other as subsets, so they form an antichain. The interval $[X, Y]$ in the Boolean consists of the $S \subseteq S_n$ for which $X \subseteq S \subseteq Y$; hence

$$|[X, Y]| = \begin{cases} 2^{|Y|-|X|} & \text{if } X \subseteq Y, \\ 0 & \text{otherwise.} \end{cases}$$

In Figure 1.7 we have drawn the Hasse diagram for the Boolean and $n = 3$. The horizontal levels of this diagram consist of antichains, more precisely, of $C^k(S_n)$; in general, the levels of the Hasse diagram are determined as the subsets whose elements have shortest paths of identical lengths to the minimal element.

Natural numbers, ordered by divisibility: We consider on the set $\mathbf{N} = \{1, 2, \dots\}$ of natural numbers the binary relation $|$ defined according to the rule: for $x, y \in \mathbf{N}$, $x \mid y \Leftrightarrow x$ divides y without remainder.

This binary relation is a partial order, with least element $1 \in \mathbf{N}$, since every integer can be divided by 1 without remainder. It is obvious that the set of prime numbers is an antichain under this relation, and that all prime numbers are atoms. If $x \mid y$, then $|[x, y]|$ is the number of divisors of y that are multiples of x, and although $\mathbf{N}$ itself is infinite, $(\mathbf{N}, \mid)$ is locally finite.

It turns out that the partial orders $\subseteq$ and $|$ are tightly related to each other. To investigate this interaction, we first consider the Hasse diagram of $(\mathbf{N}, \mid)$, depicted in Figure 1.8. This diagram makes clear that the least element is 1, that the atoms

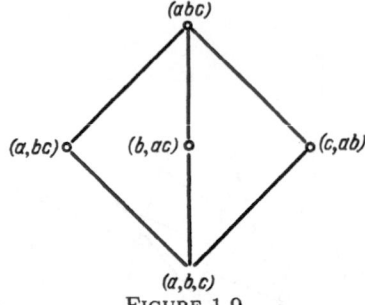

$$\text{FIGURE 1.9}$$

are the prime numbers, that the next level consists of the numbers representable as a product of two prime numbers, that the next level consists of the numbers representable as a product of three prime numbers, etc. Considering the subset $\{1, 2, 3, 5, 6, 10, 15, 30\}$ as being ordered by divisibility $|$, we become convinced of the fact that its Hasse diagram coincides precisely with the Hasse diagram of the Boolean $(\mathcal{P}(S_3), \subseteq)$.

An even tighter relation with numbers ordered by divisibility is possessed by the

Boolean of a multiset, ordered by inclusion: Let A be the finite multiset with basis $S(A) = \{a_1, a_2, \ldots, a_n\}$ and first specification $[k_1, k_2, \ldots, k_r]$, i.e. the element a_i occurs k_i times in A, $i = 1, 2, \ldots, r$. Then A has cardinality $|A| = \sum_{i=1}^{r} k_i = n$, so that

$$A = \sum_{a \in A} C(a) = \sum_{a \in S(A)} k_A(a)C(a) = \sum_{i=1}^{r} k_i C(a_i).$$

We introduce on the Boolean $\mathcal{P}(A)$ a binary relation $\subseteq$ according to the rule: for elements $X, Y \in \mathcal{P}(A)$, $X \subseteq Y \leftrightarrow X$ is a submultiset of the multiset Y. The relation of this Boolean with numbers ordered by divisibility is based on the following simple fact. If $p_1, \ldots, p_r$ are distinct prime numbers and $M = p_1^{k_1} p_2^{k_2} \ldots p_r^{k_r}$, then the partially ordered (by divisibility) set of natural divisors of M has the same Hasse diagram as $\mathcal{P}(A)$.

Bellean, ordered by subpartitioning: The above introduced procedure of splitting a partition of a set determines a partial order relation on the Bellean. We say that a partition is 'larger' than another if the second can be obtained from the first by splitting. We denote this relation by the sign $\rightarrow$. E.g., if $X = \{a, b, c, d\}$, then $(ab, cd) \rightarrow (a, b, c, d)$, since $\{a, b\} = \{a\} \cup \{b\}$, but $(ac, bd) \nrightarrow (a, b, cd)$, since the union of the blocks of the second partition cannot be obtained from blocks of the first. This partially ordered set also has least and largest elements: (a, b, c, d) and $(abcd)$, respectively. For the three-element set $\{a, b, c\}$ the Hasse diagram of the Bellean ordered by splitting is depicted in Figure 1.9.

The lexicographical order on sequences is given by the rule: $(x_1, x_2, \ldots, x_n, \ldots) \succcurlyeq (y_1, y_2, \ldots, y_n, \ldots)$ if and only if either $x_i = y_i$ for all i, or there is a natural number $i \in \mathbf{N}$ for which $x_i > y_i$ and $x_j = y_j$ for $j < i$.

This definition is correct also for finite sequences of the same length. The lexicographical order is used for ordering words in dictionaries, with the condition that if a word is shorter than another, then the last, missing, letters of the shorter word are weighted as maximal components; e.g., 'an' $\succcurlyeq$ 'and'. In this way we can use this order also for sequences of distinct lengths.

Difference, as well as equivalence of various objects, can also be conveniently characterized by binary relations.

2.7. Equivalence relations. We say that an *equivalence relation* R is given on a set X if $R \subseteq X^{(2)}$ and R is reflexive, symmetric, and transitive. An equivalence relation on a set X induces a partition of X; its blocks are called *equivalence classes*. E.g., equality ($=$) is an equivalence relation, and partitions each set into singleton (i.e. one-element) equivalence classes, and each multiset into equivalence classes consisting of copies of a single element, as many as the multiplicities of these elements in the multiset; thus, the volumes of the equivalence classes coincides in this case with the first specification of the multiset. For an arbitrary equivalence relation two elements in a single equivalence class are mutually equivalent (pairwise comparable), and elements in different equivalence classes are not equivalent (incomparable). Hence, every partition of a set uniquely determines an equivalence relation on this set.

The *quotient set* by a given equivalence relation R is the set of all equivalence classes of R. As an example, suppose an equivalence relation on the set of natural numbers is given by the rule: xRy if x and y have the same parity. This gives two equivalence classes, of the odd and the even numbers, so that the quotient set consists of two elements.

3. Binary functions on ordered sets*

In this Section we let $(P, \leq)$ be an ordered set (§2.6), and $\leq$ a reflexive binary relation (§2.5), also called an *order*. A closed chain (§2.6) in P is called a *cycle*. A chain is called an *f-chain* if it contains either a zero or a unit (regarded as an ordered set). The ordered set $(P, \leq)$ is called *acyclic* if it has no cycles, *locally acyclic* if all (closed) intervals in it are acyclic, and *f-acyclic* if it has no f-chains with cycles.

3.1. Binary functions on $(P, \leq)$. We consider real-valued functions only. Let $A(P)$ be the set of such functions of two variables, defined on P^2. If a function $f \in A(P)$ is associated with the binary relation $\leq$ on P, then f will be called a *binary function*. The simplest examples of binary functions are the *Kronecker delta*

$$\delta(x, y) = \chi_{\{x=y\}} \qquad \forall x, y \in P,$$

the *order function*

$$\zeta(x, y) = \chi_{\{x \leq y\}} \qquad \forall x, y \in P,$$

also called the *zeta function* of the set $(P, \leq)$, and the *strict order function*

$$\eta(x, y) = \zeta(x, y) - \delta(x, y)$$

(here, χ denotes the indicator function of the set occurring as its subscript).

The set $AI(P) = A_\zeta(P, \leq)$ of all binary functions on $(P, \leq)$ is defined as the subset of functions in $A(P)$ that admit a representation

$$f(x, y) = f(x, y)\zeta(x, y) \qquad \forall x, y \in P, \tag{1.1}$$

1* This Section is adapted from [74].

i.e. f assumes nonzero values only for pairs $x \leq y$ or on closed intervals (cf. §2.6), and is zero otherwise. It is easy to see that the functions δ, ζ, and η satisfy (1.1).

Note that if in (1.1) the function ζ is replaced by a fixed function $\lambda(x, y)$ for which $\lambda(x, x) = 1$ for all $x \in P$, i.e. if one considers the set of functions $A_\lambda(P, \leq) \subseteq A(P)$ that are nonzero only on the pairs x, y for which $\lambda(x, y) = 1$, then this is equivalent to considering the set $AI(P)$ relative to the following, new, order $\preceq$ on P:

$$x \preceq y \Leftrightarrow \lambda(x, y) = 1.$$

Thus, $A_\lambda(P, \leq) = A_\zeta(P, \preceq)$.

Generally speaking, we need not require the order to be reflexive, and dropping this requirement does not lead to serious changes of the following results, although its satisfaction enables us to avoid at certain places awkward calculations. This is related to the fact that there are in fact two binary relations defined on P, the order $\leq$ and the equivalence relation of 'equality' $(=)$, while the Kronecker delta, which plays a major role in the following results, is defined as a function of the second relation only (see [13], [57], [94], [179], [194]).

3.2. Operations on functions. In $A(P)$ we first introduce two simple operations:

scalar multiplication: if $\alpha \in \mathbf{R}^1$ and $f \in A(P)$, then $\alpha \cdot f(x, y) = \alpha f(x, y)$.
addition: if $f, g \in A(P)$, then $(f + g)(x, y) = f(x, y) + g(x, y)$.

It is easy to see that $A(P)$ is closed under these operations. For a locally finite P we introduce in $AI(P)$ the following general operation, whose modifications will be used throughout this Section.

K-**convolution:** if $f, g \in AI(P)$, then

$$f * g = \sum_{z \in P} f(x, z) K(x, z, y) g(z, y),$$

where $K(x, z, y)$ is a function (from P^3 into $\mathbf{R}$), called the *kernel* of the convolution.

Since $f, g \in AI(P)$, we have

$$f * g(x, y) = \sum_z f(x, z) \zeta(x, z) K(x, z, y) \zeta(z, y) g(z, y) =$$

$$= \sum_{x \leq z \leq y} f(x, z) K(x, z, y) g(z, y).$$

Since P is locally finite, this shows that convolution is well defined. A kernel is called *correct* if if it admits a representation

$$K(x, z, y) = \zeta(x, z) K(x, z, y) \zeta(z, y) \qquad \forall x, y, z \in P,$$

i.e. if K takes nonzero values on chains $x \leq z \leq y$ only.[2] The convolution operation allows us to introduce the concepts of unit and inverse function. Namely, a (right) *unit function* with respect to the K-convolution is a function $e \in AI(P)$ that satisfies

$$f * e(x, y) = f(x, y) \qquad \forall x, y \in P,$$

for every function $f \in AI(P)$.

---

[2] It is obvious that a correct kernel makes it possible to introduce the operation $*$ in all of $A(P)$.

inverse function: if $f \in AI(P)$, then a (right) *inverse function* of f is a function $g = f^{-1}$ satisfying

$$f * g(x,y) = e(x,y) \qquad \forall x,y \in P,$$

where e is some unit function. In particular, if the unit function is taken to mean the delta function, then a function inverse to the zeta function is called a *Möbius function*, denoted by $\mu(x,y)$ (see [57], [94], [179], [194]).

3.3. Stable kernels. It is obvious that $AI(P)$ is not always closed under the operations of convolution and taking the inverse. We now try to choose convolutions for which

a) $AI(P)$ is closed under convolution;
b) the Kronecker delta is a unit function;
c) there is a Möbius function.

Convolutions (and their kernels) satisfying a)–c) are called *stable on P*. If there is a Möbius function belonging to $AI(P)$, such convolutions (and their kernels) are called *Möbius stable on P*.

A criterion for a) to hold is given by

PROPOSITION 1.1. *The set $AI(P)$ is closed under K-convolution if and only if*

$$K(x,z,y) = K(x,z,y)\zeta(x,y) \qquad \forall x,z,y \in P, \quad z \in [x,y]. \tag{1.2}$$

PROOF. *Sufficiency.* We have

$$f * g(x,y) = \sum_{x \le z \le y} f(x,z)K(x,z,y)g(z,y) =$$

$$= \sum_{x \le z \le y} f(x,z)K(x,z,y)\zeta(x,y)g(z,y) =$$

$$= \zeta(x,y)\sum_{z} f(x,z)K(x,z,y)g(z,y) = \zeta(x,y)f * g(x,y).$$

Necessity. Assuming that $AI(P)$ is closed, we find that if

$$K(x_0,z_0,y_0) \ne K(x_0,z_0,y_0)\zeta(x_0,y_0), \qquad z_0 \in [x_0,y_0],$$

then $K(x_0,z_0,y_0) \ne 0$ and $1 \ne \zeta(x_0,y_0)$, i.e. $x_0 \not\le y_0$. But then the functions $f_0(x,y) = \delta(x,x_0)\delta(y,z_0)$ and $g_0(x,y) = \delta(x,z_0)\delta(y,y_0)$ on the one hand belong to $AI(P)$, and on the other hand their convolution is nonzero at (x_0,y_0), since

$$f_0 * g_0(x_0,y_0) = \sum_{z} \delta(x_0,x_0)\delta(z,z_0)K(x_0,z,y_0)\delta(z,z_0)\delta(y_0,y_0) =$$

$$= \sum_{z} \delta(z,z_0)K(x_0,z,y_0) = K(x_0,z_0,y_0) \ne 0.$$

This contradicts the fact that $AI(P)$ is closed. □

COROLLARY 1.1. *If the order is also transitive, the set $AI(P)$ is closed under any correct convolution.*

PROOF. In fact, starting from the correctness of the kernel, the transitivity implies that

$$K(x, z, y) = K(x, z, y)\zeta(x, z)\zeta(z, y) \le K(x, z, y)\zeta(x, y).$$

Since $1 \ge \zeta(x, y)$ we have

$$K(x, z, y) \ge K(x, z, y)\zeta(x, y),$$

which means that

$$K(x, z, y) = K(x, z, y)\zeta(x, y). \qquad \square$$

Condition b) imposes quite severe restrictions on the nature of the kernel.

PROPOSITION 1.2. *The delta function is a unit function under K-convolution with kernel $K(x, z, y)$ if and only if*

$$K(x, y, y) = 1 \qquad \forall x, y \in P, \quad x \le y. \tag{1.3}$$

PROOF. *Sufficiency.* If $f \in AI(P)$, then

$$f * \delta(x, y) = \sum_z f(x, z)K(x, z, y)\delta(z, y) =$$

$$= \zeta(x, y)K(x, y, y)f(x, y) = \zeta(x, y)f(x, y) = f(x, y).$$

Necessity. Assume that δ is a unit function. If $K(x_0, y_0, y_0) \ne 1$ for $x_0 \le y_0$, then $f_0(x, y) = \delta(x, x_0)\delta(y, y_0)$ belongs, on the one hand, to $AI(P)$, and on the other hand its convolution with the delta function is not equal to it at (x_0, y_0), since

$$f * \delta(x_0, y_0) = \sum_z \delta(x_0, x_0)\delta(z, y_0)K(x_0, z, y_0)\delta(z, y_0) =$$

$$= \sum_z \delta(z, y_0)K(x_0, z, y_0) = K(x_0, y_0, y_0) \ne 1 = f_0(x_0, y_0),$$

contradicting the assumption. $\square$

In a completely similar way can we demonstrate that delta is a left unit if and only if

$$K(x, x, y) = 1 \qquad \forall x, y \in P, \quad x \le y. \tag{1.4}$$

PROPOSITION 1.3. *The operation of K-convolution is associative if and only if*

$$K(x, v, y)K(v, z, y) = K(x, z, y)K(x, v, z), \qquad x \le v \le z \le y.$$

PROOF. *Sufficiency.* Let f, g, h be binary functions. Then

$$f *_K (g *_K h)(x, y) =$$

$$= \sum_{v \in P} \sum_{z \in P} f(x, v)g(v, z)h(z, y)K(x, v, y)K(v, z, y) =$$

$$= \sum_{x \le v \le z \le y} f(x, v)g(v, z)h(z, y)K(x, v, y)K(v, z, y) =$$

$$= \sum_{x \le v \le z \le y} f(x, v)g(v, z)h(z, y)K(x, z, y)K(x, v, z) =$$

$$= \sum_{v \in P} \sum_{z \in P} f(x, v)g(v, z)h(z, y)K(x, z, y)K(x, v, z) =$$

$$= (f *_K g) *_K h(x, y).$$

Necessity. If for $x_0 \leq v_0 \leq z_0 \leq y_0$ the condition of the proposition does not hold, then for the binary functions $f_0(x,y) = \delta(x,x_0)\delta(y,v_0)$, $g_0(x,y) = \delta(x,v_0)\delta(y,z_0)$, $h_0(x,y) = \delta(x,z_0)\delta(y,y_0)$ we have

$$f_0 *_K (g_0 *_K h_0)(x_0,y_0) = K(x_0,v_0,y_0)K(v_0,z_0,y_0) \neq$$
$$\neq K(x_0,z_0,y_0)K(x_0,v_0,z_0) = (f_0 *_K g_0) *_K h_0(x_0,y_0).$$

This contradicts the associativity of the kernel. $\square$

In particular, for concrete kernels this criterion takes a lattice character.

COROLLARY 1.2. *If R is reflexive and transitive, then the set of binary functions is closed under any correct convolution.*

COROLLARY 1.3 ([134]). *If P is finite, then $AI1(P,R)$ is a subalgebra of the algebra $A1(P)$ if and only if R is transitive.*

COROLLARY 1.4. *If R is reflexive and f is acyclic, then the set $AIC(P,R)$ is an algebra with unit δ (in general, not associative).*

COROLLARY 1.5. *If R is reflexive and f is acyclic, then the algebra $AIC(P,R)$ is associative if and only if every f-chain is linearly ordered in (P,R).*

Sufficient conditions for Möbius stability in the class of sets with a reflexive antisymmetric order are given by the following lemma.

LEMMA 1.1 (INVERSE FUNCTION LEMMA). *Suppose $(P,\leq)$ is a locally finite f-acyclic ordered set with a reflexive antisymmetric order. Assume that the kernel satisfies the conditions (1.2) and (1.3), and that*

$$K(x,x,y) \neq 0 \qquad \forall x,y \in P, \quad x \leq y. \tag{1.5}$$

Then a function $f \in AI(P)$ has an inverse function $f^{-1} \in AI(P)$ (relative to the delta function) if and only if $f(x,x) \neq 0 \; \forall x \in P$.

The proof repeats almost verbatim that of Lemma 2.2.1 in [94]. It is carried out by induction, and follows from the possibility of defining f^{-1} recursively as

$$f^{-1}(x,y) = \frac{\delta(x,y)}{f(x,y)} - \sum_{\substack{x < z \leq y \\ x \leq y}} \frac{f(x,z)K(x,z,y)}{f(x,x)K(x,x,y)} f^{-1}(z,y).$$

In particular, this implies $f^{-1} \in AI(P)$. Note that the assumptions of antisymmetry and acyclicity are essential in the induction argument.

COROLLARY 1.6. *Suppose the order is reflexive and antisymmetric, and P is f-acyclic. If the kernel K on P satisfies (1.2) and has delta as left and right unit, then $f \in AI(P)$ has an inverse if and only if $f(x,x) \neq 0 \; \forall x \in P$.*

In particular, $K(x,z,y) = \zeta(x,y)$ is Möbius stable on every f-acyclic ordered set with a reflexive and transitive order. One can prove that ζ is Möbius stable even on an ordered set with a reflexive and transitive order.

It is clear that, in general, the existence of a Möbius function is equivalent to solvability of the system of equations

$$\zeta * \mu(x,y) = \delta(x,y), \qquad x,y \in P,$$

for the unknown μ. If P is finite, a precise criterion for the solvability of such a system can be expressed in terms of minors.

3.4. New kernels on ordered sets. In this Subsection we assume that $(P, \leq)$ is a set that is (at least) reflexive, transitive, and antisymmetrically ordered.

Earlier papers (see, e.g., [57], [94], [179], [194]) considered mainly the unit kernel $K \equiv 1$ on a partially ordered set. Of course, this kernel is Möbius stable on $(P, \leq)$. Effective extensions of the set of kernels have only been made for sets $(P, \leq)$ of a special kind (see, e.g., [202]). Our precise conditions for stability make it possible to introduce new kernels for general $(P, \leq)$ (but still reflexive, transitive, and antisymmetric). We now consider several of these.

Let $K(x, z, y) = |[z, y]|$ be the cardinality of the interval $[z, y]$. Clearly, (1.2), (1.3), (1.5), hold for this kernel, so that it is Möbius stable.

The Möbius function is defined by

$$\mu(x, y) = \delta(x, y) - \sum_{x < z \leq y} \frac{|[z, y]|}{|[x, y]|} \mu(z, y). \tag{1.6}$$

In particular, if we consider the set of all subsets of a finite set $S_n = \{a_1, \ldots, a_n\}$, partially ordered by inclusion, then

$$|[z, y]| = 2^{|y| - |z|} \zeta(z, y) \quad \text{and} \quad \mu(x, y) = \left(-\frac{1}{2}\right)^{|y| - |x|} \zeta(x, y). \tag{1.7}$$

Let $r(x, y)$ denote the length of the largest chain with 'lower end' x and 'upper end' y. Then the following may serve as examples of Möbius-stable kernels on a partially ordered set:

$$K(x, z, y) = r(z, y) + \zeta(z, y); \tag{1.8}$$

$$K(x, z, y) = q^{r(z, y)}, \qquad q \in \mathbf{N}; \tag{1.9}$$

$$K(x, z, y) = \binom{|[x, y]|}{|[z, y]| - 1}; \tag{1.10}$$

$$K(x, z, y) = \binom{|[x, y]|}{|[x, z]| \cdot |[z, y]|}. \tag{1.11}$$

In particular, the enumeration and classification of stable (Möbius-stable) kernels makes it possible to count and classify ordinary combinatorial identities (and their inverses).

3.5. The inversion principle. Suppose that there is defined of the set $AI(P)$ of all binary functions a K-convolution with kernel $K(x, z, y)$.

THEOREM 1.1. *Let $(P, \leq)$ be a locally finite, reflexively ordered set with zero 0_P. Suppose that we have functions $f, g, \lambda, \kappa \in A(P)$, defined on P^2, satisfying*

$$g(y, x) = \sum_{z \leq y} f(z, x) K(z, y, x) \qquad \forall x, y \in P, \tag{1.12}$$

$$\zeta * \zeta \lambda(x, y) = \delta(x, y) \kappa(x, y) \qquad \forall x, y \in P. \tag{1.13}$$

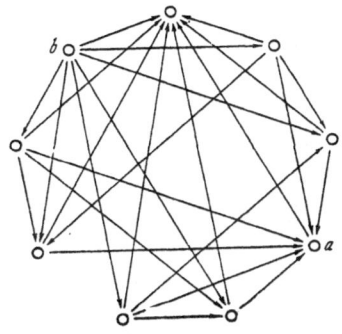 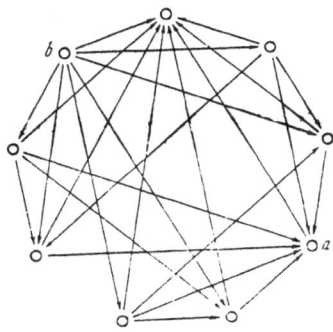

Then

$$f(x,x)\kappa(x,x) = \sum_{y\le x} g(y,x)\lambda(y,x). \qquad (1.14)$$

PROOF. Since $0_P \in P$, the local finiteness of $(P,\le)$ shows that all sums below are well defined:

$$\sum_{y\le x^y} g(y,x)\lambda(y,x) = \sum_{y\le x^y}\left(\sum_{z\le y^z} f(z,x)K(z,y,x)\right)\lambda(y,x) =$$

$$= \sum_{y\le x^y}\sum_{z} f(z,x)\zeta(z,y)K(z,y,x)\lambda(y,x) =$$

$$= \sum_{z} f(z,x)\sum_{y}\zeta(z,y)K(z,y,x)\zeta(y,x)\lambda(y,x) =$$

$$= \sum_{z} f(z,x)(\zeta * \zeta\lambda(z,x)) =$$

$$= \sum_{z} f(z,x)\delta(z,x)\kappa(z,x) = f(x,x)\kappa(x,x),$$

as required. $\square$

It is clear that the existence of a zero can be replaced by the finiteness of sums of the types (1.12) and (1.14).

If $(P,\le)$ is moreover antisymmetrically ordered, then $\kappa(x,x) = K(x,x,x)\lambda(x,x)$, and if δ is a unit function for K-convolution, then $\kappa(x,x) = \lambda(x,x)$.

If $\kappa \equiv 1$, then $\zeta\lambda = \mu$, whereas if $\lambda \in AI(P)$, then $\zeta\lambda = \lambda = \mu$. If K is a Möbius-stable kernel, then for $\kappa \equiv 1$ we can always choose a $\lambda \in AI(P)$ satisfying (1.13); in this case the requirement of Möbius stability may be replaced by that of correctness and stability of the kernel K. Note also that for $\kappa \equiv 1$ and $(P,\le)$ at least antisymmetrically ordered there is a reflexive, antisymmetrically ordered set for which (1.13) holds for $K \equiv 1$ and 'ordinary' μ:

$$\mu(x,y) = \begin{cases} 1, & x = y, \\ -\sum_{x<z\le y}\mu(z,y), & x < y, \\ 0, & x \not\le y. \end{cases} \qquad (1.15)$$

An example of such a reflexive, antisymmetrically ordered set without cycles is shown in Figure 3.5 (where an arrow means that strict inequality $<$ holds). Clearly, this example is not transitive, but a simple verification shows that (1.13) holds. By adding an arrow from a to b we obtain a set with cycles which still satisfies (1.13).

Finally, if we take the order antisymmetric and transitive, $K \equiv 1$, $\kappa = 1$, and $f(z, x) = f(z)$, then $\lambda = \mu \in AI(P)$ and Theorem 1.1 becomes the ordinary Möbius inversion principle for partially ordered sets (see [57], [94], [179], [194]).

3.6. Calculation of μ for antisymmetric orders. Here $(P, \leq)$ will be taken locally finite, f-acyclic, and reflexively and antisymmetrically ordered. On $AI(P)$ we define the convolution with kernel $K(x, z, y) = \zeta(x, y)$. Then $AI(P)$, endowed with the operations of addition, scalar multiplication, and ζ-convolution, is an algebra, with unit element δ.

Let $\tau_0(x, k, y)$ be the number of chains of length k with initial element y and zero element x in the set $(P, \leq)$. A method for computing the values of a Möbius function, not relying on recursion, is given by the following theorem.

THEOREM 1.2. *Let $(P, \leq)$ be a locally finite, f-acyclic ordered set with a reflexive and antisymmetric order. Then*

$$\mu(x, y) = \sum_{k \geq 0} (-1)^k \tau_0(x, k, y) \qquad \forall x, y \in P. \tag{1.16}$$

PROOF. Introduce the notation

$$f^{(n)} = \underbrace{(\ldots ((f * f) * f) * \cdots * f) * f}_{n},$$

where $f^{(0)} = \delta$. Since $AI(P)$ is an algebra, it immediately follows that

$$\mu = \zeta^{-1} = (\delta + \eta)^{-1} = \delta - \eta^{(1)} + \eta^{(2)} - \eta^{(3)} + \ldots,$$

i.e.

$$\mu(x, y) = \sum_{k \geq 0} (-1)^k \eta^{(k)}(x, y) \qquad \forall x, y \in P. \tag{1.17}$$

Since we can easily check that

$$\eta^{(k)}(x, y) = \tau_0(x, k, y) \qquad (k \geq 0),$$

we obtain the required result. □

Note that (1.16) gives only the value of a left Möbius function. The similar formula for a right Möbius function has the form

$$\mu(x, y) = \sum_{k \geq 0} (-1)^k \tau_1(x, k, y), \tag{1.16'}$$

where $\tau_1(x, k, y)$ is the number of chains in P of length k with largest element y and terminal element x.

In particular, if the order is transitive and antisymmetric, then $\tau_0 = \tau_1$, and formulas (1.16) and (1.16') coincide and give Ph. Hall's theorem (see [13], [179], [194], [203]).

Furthermore, even if the ζ-kernel is Möbius stable on a locally finite antisymmetrically ordered set $(P, \leq)$ (which need not be f-acyclic, as above), then the requirement of f-acyclicity in Theorem 1.2 is essential, since the presence of a cycle in an f-chain would imply that there are infinitely many summands in (1.16), resulting, in particular, in equalities of the type

$$\frac{1}{2} = \sum_{k=0}^{\infty}(-1)^k. \tag{1.18}$$

In fact, it suffices to compute the Möbius function $\mu(x, y)$ of the binary relation

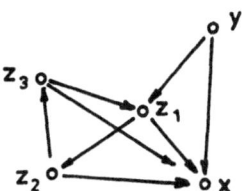

in explicit form, and to use the above-given analog of Ph. Hall's theorem, noting that there is a unique chain of length k from y to x (for each $k \geq 1$), and that $\mu(x, y) = -1/2$.

At first glance the algebras introduced in this Section may seem utmost abstract constructs. It turns out that they are far from this, and *Stanley's phenomenon* may serve as a first substantiation of this. Nowadays there is a whole theme with origin given by *Stanley's theorem*, which asserts that locally finite partially ordered sets are isomorphic if and only if their incidence algebras are isomorphic (as abstract algebras). Thus, the structure of the basement (a set with a binary relation) is uniquely determined by the corresponding algebra of functions. And this phenomenon takes place for a large class of binary relations and algebras; the main results here are due to V. Shmatkov.

Thus, the apparatus of incidence algebras may serve as a model for the actual phenomena that can be adequately described by sets with binary relations. A concrete example is given by a combinatorial medium, including such in the form of an electronic network.

Precisely the construction of algebras of functions over arbitrary binary relations crucially extends the sphere of modeling actual phenomena, which was earlier restricted by the axioms of partial order.

4. Combinatorial schemes

Combinatorial schemes are the most typical and in practice most often used combinatorial formations. The use of some combinatorial scheme is determined by the initial statement of a problem and by the choice of a method for solving it.

Among the simplest combinatorial schemes, the following two are used most often: combination and arrangement. Both these schemes are easy to comprehend by invoking yet another natural notion, that of *selection*, more precisely as an action of choosing some objects from a given collection. So,

r-combination from n elements: the name given to the result of choosing r elements from n elements, not taking into account their order;

r-arrangement from n elements: the name given to the result of choosing r elements from n elements, taking into account their order.

Since we choose from a set, all n initial elements are distinct, and an r-combination is an r-element subset, while for $r = n$ an r-arrangement is a permutation of the initial elements.

Next to the natural notion of selection, a systematization of the simplest combinatorial schemes is also provided, and here we propose the

listing scheme: The simplest combinatorial unification of certain objects is by listing them, i.e. representing these objects by symbols: elements of a list (usually written in row form $(a_1, a_2, \ldots, a_n, \ldots)$). E.g., a set given as a list of its elements is such. The set consists of distinct elements, whose order is immaterial; combination by simple set attributes, including being different and the sequence order, is the methodological basis of combinatorics, and induces all the most simple combinatorial schemes for lists:

lists of distinct elements, whose order is immaterial: sets, combinations. E.g., $\{1,2\}, \{1,3\}, \{2,3\}$ are all 2-combinations of the set $\{1,2,3\}$. The number of all r-combinations from an n-element set equals the binomial coefficient $\binom{n}{r} = n!/(r!\,(n-r)!)$.

lists of not necessarily distinct elements, whose order is immaterial: multisets, combinations with repetitions, selection, totalities, families. E.g., $\{1,1\}, \{1,2\}, \{1,3\}, \{2,2\}, \{2,3\}, \{3,3\}$ are all 2-combinations with repetitions from the three element set $\{1,2,3\}$ or, in other words, all two-element submultisets of the multiset $\{1^2, 2^2, 3^2\}$. The number of all r-combinations with repetitions (without restrictions on the number of repetitions) from n elements is $\binom{n+r-1}{r}$.

lists of distinct elements, whose order is essential: arrangements, permutations. E.g., $\{1,2\}, \{1,3\}, \{2,3\}, \{2,1\}, \{3,1\}, \{3,2\}$ are all 2-arrangements from the three element set $\{1,2,3\}$; while $(1,2,3), (1,3,2), (2,3,1), (3,1,2), (3,1,2)$ are all permutations of the set $\{1,2,3\}$. The number of all r-arrangements from n elements is $r!\binom{n}{r}$, hence the number of permutations of an n-element set is $n!$.

lists of not necessarily distinct elements, whose order is essential: arrangements with repetitions, permutations with repetitions, sequences, vectors, tuples. E.g., $(1,1), (1,2), (2,1), (1,3), (3,1), (2,2), (3,3), (2,3), (3,2)$ are all 2-arrangements with repetitions from the three elements $\{1,2,3\}$. The number of all r-arrangements with repetitions (without restrictions on the number of repetitions) from n elements is n^r.

types of lists according to: being distinct / being ordered	without repetitions	with arbitrary many repetitions	arrangement / type / pellet
unordered	$\binom{n}{k}$	$\binom{n+k-1}{k}$	not distinct
ordered	$\frac{n!}{(n-k)!}$	n^k	distinct
sample type / selection	without return	with return	

Table 1.1.

This systematization allows us to arrange the simplest combinatorial schemes and their most important numerical characteristics as in Table 1.1. This table immediately shows the equivalence of certain combinatorial schemes:

the listing scheme of k-elements from n objects;
the scheme of arranging k pellets over n distinct boxes;
the urn scheme of choosing k balls from n distinct urns.

types / boxes / particles	distinct	not distinct
distinct: ordered in boxes and blocks	$n=3,\ k=2$ $(a,bc)(bc,a)(a,cb)$ $(cb,a)(b,ac)(ac,b)$ $(b,ca)(ca,b)(c,ab)$ $(ab,c)(c,ba)(ba,c)$	$n=3,\ k=2$ $(a,bc)(a,cb)$ $(b,ac)(b,ca)$ $(c,ab)(c,ba)$
distinct: unordered in boxes and blocks	$n=4,\ k=2$ $(a,bcd)(bcd,a)(b,acd)$ $(acd,b)(c,abd)(abd,c)$ $(d,abc)(abc,d)(ab,cd)$ $(cd,ab)(ac,bd)(bd,ac)$ $(ad,bc)(bc,ad)$	$n=4,\ k=2$ $(a,bcd)(ab,cd)$ $(b,acd)(ac,bd)$ $(c,abd)(ad,bc)$ (d,abc)
not distinct	$n=6,\ k=3$ $(4,1,1)(1,4,1)(1,1,4)$ $(3,2,1)(3,1,2)(1,3,2)$ $(2,3,1)(2,1,3)(1,2,3)$ $(2,2,2)$	$n=6,\ k=3$ $(4,1,1)$ $(3,2,1)$ $(2,2,2)$
objects / blocks / types	ordered	unordered

Table 1.2

The numerical values in the table entries denote the amounts of corresponding combinatorial formations.

Table 1.2 illustrates the following schemes:

the scheme of arranging n particles over k boxes without empty boxes;

the scheme of partitioning n objects into k blocks;

the scheme of representing a natural number n as a sum of k natural numbers.

So, these schemes (listing, selection, arrangement, partitioning, urn) have allowed us to systematize all elementary combinatorial formations. However, not for every problem do the existing combinatorial schemes ensure a unified solution, leaving thus unrealized the Leibniz dream of creating an all-embracing combinatorial scheme, although the phrase 'general combinatorial scheme' has already been 'in use'. In essence, this combinatorial scheme is represented in Tables 1.1 and 1.2, with the single difference that the concrete contents of entries of the latter must be replaced by general 'mechanisms' modelling the initial combinatorial conditions and types, and by general means for calculating the necessary numerical characteristics (mainly by the method of generating functions or other enumeration methods). For a more detailed acquaintance with the general combinatorial scheme see [60], [62].

Thus, the notion of combinatorial scheme includes a practical, positive attempt to a unified approach to some or other circle of combinatorial problems.

5. Combinatorial problems and their complexity

With the appearance of powerful computers the role of mathematical disciplines in which discrete systems are studied has become essentially larger. Presently, the level of scientific-technological processes and the problems posed in practice have produced most profound demands on the methods of investigating such discrete systems as graphs, block-designs, matrices, electrical circuits, transportation flows, information flows, manufacture organizational systems, logical schemes, and many others. Despite the apparent variety and specific natures of these discrete systems, in their basics these systems have much in common. We can unify their mathematical models and the methods for solving them in individual types, and here we consider the main such types.

5.1. Types of combinatorial problems. In the most general case the conditions of any problem are (the problem statement is) determined by a list of initial parameters or initial data and a description of the required result or of the properties the answer has to satisfy.

Combinatorial problems are problems with discrete variables only. As a rule, the initial data of such problems are combinatorial objects: numbers, vectors, sets, graphs, families of sets, number partitions, etc. Depending on the description of the result looked for, the solutions of combinatorial problems may be either the answer 'yes' or 'no', or some object from a finite (infinite) set, i.e. a set of elements with given properties, and also graphs, numbers, partitions, etc. The differences in the descriptions of the required result of solutions combinatorial problems can be grouped together into individual types, characterized by common properties of the final results. Here we will consider the basic types of combinatorial problems.

Counting problems are combinatorial problems whose solution consists of the number of elements of a set having some property, or a family of properties, given in the problem statement. We give some examples of such problem statements.

EXAMPLE 1.1. For a given partition $(n_1, \ldots, n_r)$ of a number n, determine the amount of partitions of a number $k \leq n$ packed in it.

EXAMPLE 1.2. For a given graph G, determine the number of simple chains of length three in it.

EXAMPLE 1.3. In how many ways can we distribute n elements over n boxes (one in each box) such that no ith element falls into box i?

EXAMPLE 1.4. In how many ways can we make an unordered selection of 7 balls from an urn with 11 distinct balls, without replacement of balls (with replacement)?

Enumeration problems are combinatorial problems in which the solutions are objects with given properties (numbers, sets, graphs, number partitions, partitions of sets, etc.). The following are examples of such problem statements.

EXAMPLE 1.5. List the partitions of a number k that are packed in a partition $(n_1, \ldots, n_r)$ of a number $n \geq k$.

EXAMPLE 1.6. List all paths of a graph G whose lengths do not exceed a given length.

EXAMPLE 1.7. List the permutations of the five letters A, B, C, D, E in which no two consonants are placed next to each other.

We note that in certain handbooks enumeration problems are defined as counting problems. However, for such a definition the problem of finding the whole set of solutions is not singled out as an individual type. Therefore we use the solution proposed in [52], in which problems whose required result is the whole set of solutions are singled out as an individual type of enumeration problem, while finding the amount of admissible solutions is defined as being a counting problem.

Recognition problems are combinatorial problems with as solution the answer 'yes' or 'no'. This type of problem is fundamental in our investigation, hence we consider in more detail some examples of stating such problems. Presently, intensive studies are devoted to such problems, and many such problems have been given a proper name, characterizing the required solution result.

EXAMPLE 1.8 (PACKABILITY OF PARTITIONS PROBLEM). For a given pair of partitions of numbers k and n, determine whether the partition $(k_1, \ldots, k_t)$ is packed in $(n_1, \ldots, n_r)$.

EXAMPLE 1.9 (HAMILTONIAN PATH). For a given directed graph $G = (V, A)$, determine whether G has a *Hamiltonian path*, i.e. a directed path passing through each vertex in V exactly once.

EXAMPLE 1.10 (GRAPH COLORING). For a given graph $G = (V, E)$ and integer k, determine whether there is a map $\phi \colon V \to \{1, \ldots, k\}$ such that $[v, u] \in E$ implies $\phi(v) \neq \phi(u)$.

EXAMPLE 1.11 (PATH IN A DIRECTED GRAPH). For a given graph $G = (V, A)$ and subsets $S, T \in V$, determine whether G contains a path from some vertex in S to a vertex in T.

EXAMPLE 1.12 (MAXIMAL PAIRING). For a given graph $G = (V, E)$ and integer k, determine whether G contains a set M of k edges such that no two edges in M have a common vertex in M.

EXAMPLE 1.13 (GRAPH CLIQUE). For a graph $G = (V, E)$ and integer k, determine whether G contains a *clique of dimension* k, i.e. a totally connected subset of V.

EXAMPLE 1.14 (TRAVELLING SALESMAN PROBLEM). For a finite set $C = (c_1, \ldots, c_m)$ of 'cities', given 'distances' $d(c_i, c_j) \in \mathbf{Z}^+$ between each pair $c_i, c_j \in C$ and given integer $B \in \mathbf{Z}^+$, determine whether there is a 'route' passing through all cities in C and whose length does not exceed B. Here $\mathbf{Z}^+$ is the set of positive integers.

EXAMPLE 1.15 (INTEGER LINEAR PROGRAMMING PROBLEM). For an integral $(m \cdot n)$ matrix A and an integral m-vector b, determine whether there is an integral n-vector x such that $Ax = b$, $x \geq 0$.

EXAMPLE 1.16 (3-DIMENSIONAL COMBINATION). For three given sets U, V, W of the same cardinality and a given subset T of $U \cdot V \cdot W$, determine whether T contains a subset M such that $|M| = |U|$ and such that if (u, v, w) and (u', v', w') are different triplets in M, then $u \neq u'$, $v \neq v'$, $w \neq w'$.

EXAMPLE 1.17 (INTEGRAL KNAPSACK). For given integers c_j, $j = 1, \ldots, n$, and k, determine whether there are integers $x_j \in \{0, 1\}$, $j = 1, \ldots, n$, such that $\sum_{d=1}^{n} c_d x_d = k$.

EXAMPLE 1.18 (PARTITIONING). For given numbers $a_1, \ldots, a_n$, does there exists a subset $S \subseteq \{1, \ldots, n\}$ such that $\sum_{j \in S} a_j = \sum_{j \notin S} a_j$?

EXAMPLE 1.19 (MAXIMAL GRAPH CUTTING). For a given graph $G = (V, E)$ and integer k, determine whether there is a partition of V into subsets V_1, V_2, $V_i \neq \emptyset$, $i = 1, 2$, such that E contains at most k edges joining V_1 with V_2.

EXAMPLE 1.20 (SCHEDULING IN A MULTIPROCESSING SYSTEM). For a given set A of 'data', with execution lengths $l(a) \in \mathbf{Z}^+$, with number of processors $m \in \mathbf{Z}^+$, and with discrete list $D \in \mathbf{Z}^+$ of execution of these data, determine whether there exists a partition $A = A_1 \cup \cdots \cup A_m$ into disjoint subsets such that

$$\max_{a \in A_i} \{ \sum l(a) : 1 \leq i \leq m \} \leq D.$$

EXAMPLE 1.21 (VERTEX COVERING). For a given graph $G = (V, E)$ and integer $k \leq |V|$, determine whether G has a vertex cover of at most k elements, i.e. a subset $V' \subseteq V$ such that $|V'| \leq k$ and for each edge $\{u, v\} \in E$ at least one of u, v belongs to V'.

Mastering problems are combinatorial problems for which the final result of solving them is either an admissible solution, or the answer 'no'. In fact, a number of combinatorial problems do not have one solution, but a set of solutions, for one and the same initial data. For example, the problem of determining (constructing) a Hamiltonian cycle has as many solutions as there are Hamiltonian cycles in the graph under investigation. The finding of one of them is a result of solving a mastering problem, while if no one is found (the graph does not have such a cycle), the result is 'no'. In other words, when stating the required result, mastering problems do not have the restrictions in the conditions which we would have when searching for a single unique solution in a set of solutions. Hence, the mastering problem of packability of number partitions can be formulated as follows:

For a given partition $(n_1, \ldots, n_r)$, find a partition of rank t of a number $k \leq n$ packed in this partition.

It is not especially difficult to formulate mastering problems on the basis of the Examples 1.8–1.20 as a type of recognition problems, giving restrictions on the properties of the object looked for.

Combinatorial optimization problems are extremal combinatorial problems whose solution is an object with concrete properties, given in the conditions of the problem. Usually such problems are given as restrictions, and the solution of the problem requires one to consider the whole subset of admissible solutions. The aim of solving all imaginable combinatorial optimization problems is the construction of optimal objects, such as circuits, routes, vertex sets, partitions and lists of integers, etc. Examples of extremal combinatorial problems are problems in which we are required to find a maximum (minimum) of a function F defined on a set of systems $Q = (\pi_1, \ldots, \pi_n)$. Such systems may be partitions, combinations, various subsequences, etc. Usually, when solving such problems in general it turns out to be difficult to evaluate the relative magnitudes of $F(\pi_1)$ and $F(\pi_n)$ without directly calculating them, while the determination of $F(\pi)$ for a concrete π sometimes requires the solution of another combinatorial problem.

We consider a version of the optimization problem of packability of number partitions:

For given partitions $(n_1, \ldots, n_r)$ and $(k_1, \ldots, k_t)$, determine a version of packing $(k_1, \ldots, k_t)$ into $(n_1, \ldots, n_r)$ such that, if $n > k$, the rank of the partition of the remainder is minimal.

As for mastering problems, optimization problems can be formulated on the basis of recognition problems.

Obtaining recognition problems from optimization problems, and vice versa, is an important tool. The procedure of this reformulation is not difficult. In fact, the problem in Example 1.14 on the existence of a 'route' whose length does not exceed an integer B can be made an extremal combinatorial problem. For this it suffices to remove the restriction imposed by B, and formulate the result of solving the problem as:

Find a 'route' of minimal length which passes through all cities in C.

Exhausting information, including a couple of hundreds of similar problem statements, can be found in [21], [50].

So, we have given definitions and examples of the statement of combinatorial problems of five types: counting, enumerating, recognition, mastering, and optimization. From the viewpoint of a most general approach to the elaboration of methods for solving them, at first glance it may seem inconvenient to single out some of these as individual types, since the methods for solving them are the same. In fact, for solving the counting problem of Example 1.1 we can use the method for solving the enumeration problem in Example 1.5, i.e. enumerate all partitions of $k \leq n$ packed in $(n_1, \ldots, n_r)$ and count their number. However, we use such a solution method only because we do not know a polynomial or analytical expression allowing us to compute the required quantity (the number of packed partitions), not applying the method of mastering.

We note that also the solution of the enumeration problem on packability of partitions encounters great difficulties. This can be explained by the fact that we do not know an efficient method for solving the recognition problem on packability of partitions, as will be shown in the sequel. But not all combinatorial problems or, rather, methods for their solution can be characterized by such a state of affairs. A set of methods for solving counting and enumeration combinatorial problems has been developed; it can be found in [51], [52].

The aim of the present investigation is to develop methods for solving combinatorial recognition problems, which are at the basis of the central set-theoretical problem of all of discrete mathematics: can we exclude mastering when solving discrete problems? In other words, the discussion is about the principal possibility of finding a required solution (e.g., an optimal one), without running over all or almost all possible versions of solving the problem. This problem is not only of mathematical nature, it is also a cognitive problem. The cognitive side of the problem lies in distinguishing the class of problems for which there is no solution algorithm. As an example we may give Hilbert's tenth problem: Does there exist an algorithm which recognizes, for a given polynomial $p(x_1, \ldots, x_n)$ with integer coefficients, whether the equation $p(x_1, \ldots, x_n) = 0$ has a solution in integers. Yu.V. Matiyasevich proved in [49] that such an algorithm is principally impossible. Also, the existence has been proved of combinatorial problems for which the solution can only be obtained by exponential algorithms.

Consequently, the solution of combinatorial problems of this type must be 'carefully' approached, i.e. taking into account the possible absence of any algorithmic solution method at all, as well as the possibility of solving them by exponential algorithms only.

The lasting attempt to develop a theory of combinatorial problems, and the practice of solving such problems, led, in the 1970's, to the sudden emergence of an interest in the analysis of the computational complexity of solving such problems.

5.2. Complexity of combinatorial problems. First of all we give a definition of complexity of a problem, and give some notions on which the analysis of complexity rests.

An *algorithm* is a strict sequence of instructions (steps), making up a procedure for solving a problem posed. Without loss of generality in this definition, we consider an algorithm to be a computer program, written in a formal machine language. It is convenient to assume that an algorithm is a solution of the problem posed if it gives a valid answer on the whole domain of admissible values of the problem. For example, an algorithm $\mathcal{A}$ is not a solution of the travelling salesman problem if, as the result of its execution, it does not give a minimal route for at least one set of initial parameter specifications C, $d(c_i, c_j)$, B.

The complexity of combinatorial problems is defined in terms of the complexity of the algorithms for solving them. In fact, it is intuitively clear that if for a given problem there is no algorithm known that would give a 'fast' exact answer, then the problem can be regarded as complex. The *measure of complexity of algorithms* is taken to be their execution time, considered as a function of all the parameters figuring in the domain of applicability of the algorithm.

The *input dimension of a combinatorial problem* is the number of symbols with which the input parameters of the problem can be coded. As a rule, the code of such parameters is defined in a pre-given fixed alphabet. For example, any input data of the combinatorial problem of recognizing packability of number partitions can be coded in the alphabet $\{/, 0, 1, 2, 3, 4, 5, 6, 7, 8, 9\}$. If we pose the problem of determining the packability of the partition $(3, 2, 2, 1, 1)$ in the partition $(4, 2, 2, 2)$, then the following sequence of symbols is its input: $4/2/2/2//3/2/2/1/1$. The input dimension of this problem is 24.

The *time complexity of an algorithm* is a quantity reflecting the time spenditure in executing the algorithm with a given input dimension. It can be computed as a function which assigns to every concrete input dimension the quantity that is the maximal time of executing the algorithm for various values of the parameters in initial data of this dimension. It is clear that the execution time of an algorithm depends not only on the method of solution embodied in it. It will also depend on the type of computer on which it is implemented, and also on its implementation. To exclude such indefiniteness when estimating the time complexity of algorithms, the algorithm is expressed by a number of elementary steps (operations), to be executed by a hypothetical computer. Here it is assumed that the time of completing any elementary step is equal to the unit. Thus, the *time complexity of an algorithm* is defined as the maximal number of elementary operations needed for executing the algorithm when solving a problem with fixed input dimension on the set of possible concrete input data for this problem.

Presently there are two levels of complexity of algorithms. In their definition, we write that a function $f(x) = O(g(x))$ if there is a constant C such that for sufficiently large x the inequality $f(x) \leq Cg(x)$ holds. Here, $f(x)$ and $g(x)$ are functions defined on the set of positive integers and taking positive real values.

A *polynomial algorithm* is an algorithm whose time complexity is bounded above by a polynomial $p(r)$, where r is the input dimension of the combinatorial problem. In other words, the complexity of such an algorithm grows polynomially with the input dimension of the problem, and the *rate of growth* is defined as the degree of this polynomial.

An *exponential algorithm* is an algorithm for which there is an exponential lower bound for its time complexity.

It is clear that, when solving a single problem, various algorithms will have different time complexities. The choice of this parameter as determining factor in deciding the efficiency of an algorithm is not accidental. In fact, the time of solving extremal combinatorial problems turns out to be a critical parameter for the majority of processes and phenomena formalized by such problems.

Thus, the *complexity of a combinatorial problem* is defined as the complexity of the algorithms for solving them. We will consider certain aspects of the efficiency of applying the above given types of algorithms from this point of view.

5.3. Efficiency of algorithms for solving combinatorial problems. In the widest sense of the word, the notion of *efficiency of an algorithm* is related to the use of all computational resources necessary for its computer implementation. However, when comparing the efficiencies of algorithms it is usual to take their working time as basis. Such a comparison of algorithms can be done by different means or by various criteria. Two such means are encountered most often: according to the *mean working times* of the algorithms, or according to the *duration of their working times*. When comparing algorithms according to mean working time we take as measure the averaged value of this time on the whole domain of definition of the algorithm. When comparing by duration of working times, we evaluate the time at a point in the domain of definition at which the algorithm has to work longest of all.

The choice of a comparison measure determines the practical use of the algorithm. Therefore, since time restrictions are often a dominating factor, characterizing the fitness of some algorithm in practice, when considering methods for solving combinatorial problems our main attention will be concentrated on this kind of resource.

The characteristics of the complexities of algorithms introduced above determine to some extend the efficiency of using the algorithms. In fact, polynomial algorithms are conveniently taken to be such for which the functions of the input dimension of the problem (r) are the functions r, $r \log r$, r^2, r^3, r^5, while for exponential algorithms this function $f(r)$ has, e.g., the form 2^r, 3^r, $r^{\log r}$, $r!$. It is usual to consider polynomial algorithms to be more efficient for practical purposes. This is supported by the following arguments.

It is clear that with increasing input dimension of a problem, every polynomial algorithm for its solution is more efficient than every exponential algorithm. We prove this assertion using Table 1.3. In it we have given tentative times needed by polynomial and exponential algorithms (for various input dimensions of combinatorial problems) on a hypothetical computer realizing one step of any algorithm in 1 microsec. The distinction between the two types of algorithms is especially clear when solving problems of large dimensions. We must note that the majority of exponential algorithms are simply variants of solution by the method of complete sorting out.

n / function	10	20	30	50
$n \log n$	33 microsec	86 microsec	0.1 millisec	0.3 millisec
n^2	0.1 millisec	0.4 millisec	0.9 millisec	2.5 millisec
n^3	1 millisec	8 millisec	27 millisec	0.13 sec
n^5	0.1 sec	3.2 sec	24.3 sec	5.2 min
2^n	1 millisec	1 sec	17.9 min	35.7 yr
3^n	59 millisec	58 min	6.5 yr	$2 \cdot 10^8$ cent

Table 1.3

Another positive property of polynomial algorithms in comparison to exponential algorithms is their great 'sensitivity' to an increase in computer speed. Table 1.4 testifies of this. In it we have indicated the dimensions of problems, for various complexity functions, which can be processed by a hypothetical computer with fast speed $R = 1$ op/sec during 1 hour of functioning, and which can be processed by it if R is increased a 100 and a 1000 times. This table testifies to the fact that for polynomial algorithms an increase in computer speed gives a substantial increase in the possible dimensions of problems to be solved, while for exponential algorithms only additive increase in the possible dimensions of the problems occurs.

function	computer with high speed $R = 1$	$R = 100$	$R = 1000$
n	N_1	$100 \, N_1$	$1000 \, N_1$
n^2	N_2	$10 \, N_2$	$3.61 \, N_2$
n^3	N_3	$4.64 \, N_3$	$10 \, N_3$
n^5	N_4	$2.5 \, N_4$	$3.98 \, N_4$
2^n	N_5	$N_5 + 6.64$	$N_5 + 9.97$
3^n	N_6	$N_6 + 4.19$	$N_6 + 6.29$

Table 1.4

Yet another fact testifying to the efficiency of polynomial algorithms is the property that they are a closed class, i.e. the possibility of combining polynomial algorithms for solving partial cases of compound combinatorial problems—the combined algorithm is also polynomial.

Despite these arguments in favor of the efficiency of using polynomial algorithms, it is necessary to study also the other side of characteristics of polynomial algorithms. Is an algorithm of complexity r^{80} a practical solution of a problem? Obviously not, since the time needed for executing it even for input dimension of the problem $r = 3$, is an astronomical number, and it may turn out that some exponential algorithm is much more efficient for this input dimension. For example, it is clear from Table 1.3 that for $r \leq 20$ the algorithm with complexity 2^r is much more efficient than the algorithm with complexity r^5. Therefore, in a finite time span the efficiency of application of some algorithm must be determined by the concrete conditions of the problem to be solved. Suppose we have algorithms $\mathcal{A}_1, \mathcal{A}_2, \mathcal{A}_3, \mathcal{A}_4$, and $\mathcal{A}_5$, whose complexities are $1000 \, r$, $100 \log r$, $10 \, r^2$, r^3, and 2^r, respectively. Then $\mathcal{A}_5$ is more efficient than the others for input dimensions $2 \leq r \leq 9$, the algorithm $\mathcal{A}_3$ is such for input dimensions $10 \leq r \leq 58$, $\mathcal{A}_2$ for $59 \leq r \leq 1024$, and $\mathcal{A}_1$ for $r \geq 1024$. Moreover, there are well-known exponential algorithms recommended for use in practice. For example, the simplex method for solving linear programming problems, as well as the branch and bound method, has exponential complexity.

Here it is important to note that polynomial algorithms can be constructed only if we succeed in penetrating more deeply into the essence of the solution of a problem. Therefore, even the discovery of a polynomial algorithm of complexity r^{80} or r^{1000}, which cannot be regarded as efficient in practice, is a step towards the solution of the problems posed.

The difference, as regards algorithmic complexity, of solutions of combinatorial problems and their intensive study, which started in the 1970s, has led to a classification of all combinatorial problems in terms of complexity.

5.4. Classification of combinatorial problems in terms of complexity of their solution. Having considered two types of algorithms for solving combinatorial problems, we have naturally taken into account the fact that not for all practically posed problems there are efficient solution algorithms. Whether such algorithms exist or not for certain problems is unknown, but the search for efficient solution algorithms, which started in the 1950s, has taken one of the important places in discrete mathematics. Failures made by experts in this field of investigation have served as source for developing a theory which allows us to produce relative complexity bounds for newly arising combinatorial problems already at the stage of posing them. At the basis of this theory lies the investigation of combinatorial recognition problems which on the one hand turn out to be convenient for conducting such a kind of investigation, and on the other do not restrict the generality of the theory itself, since for each extremal combinatorial problem we can without much difficulty state a recognition problem strongly related to it. For a short exposition of the essence of this theory we start with basic notions and definitions.

The *class* $\mathcal{NP}$ of combinatorial recognition problems consists of problems having an exponential upper bound for the complexity of solving them, while the verification of the correctness of a solution is ensured by a polynomial algorithm. For example, the solution of the problem in Example 1.13 on the existence of a clique in a graph $G(V, E)$ of dimension κ may consist of successive verification of all $\binom{|V|}{\kappa}$ subsets of V having cardinality κ. It is obvious that the number of such subsets grows exponentially. However, for each set, the verification whether it is a solution (connectedness of the edges in E) can be done by a polynomial algorithm.

The *class* $\mathcal{P}$ of combinatorial recognition problems consists of the problems having a polynomial solution algorithm (which may yet be unknown). In other words, $\mathcal{P}$ consists of relatively simple combinatorial problems.

After these definitions, the question naturally rises of the interrelationships of these classes. This question arises first and foremost because it is not difficult to prove a simple relation between these two sets of combinatorial problems: namely, $\mathcal{P}$ is a subset of the problems in $\mathcal{NP}$, i.e. $\mathcal{P} \subseteq \mathcal{NP}$. However, the question whether the classes $\mathcal{P}$ and $\mathcal{NP}$ coincide or not has up till now not been answered, and it is one of the central problems in all of discrete mathematics: If $\mathcal{P} \neq \mathcal{NP}$, then what combinatorial problems constitute the set $\mathcal{NP} \setminus \mathcal{P}$, and what are their complexities? In the investigation of the relative complexity of combinatorial problems of this set, and of the class $\mathcal{NP}$ indeed, an important role is played by a concept from mathematical logic.

A *polynomial reduction of a recognition problem* A_1 to a recognition problem A_2 is a polynomial algorithm $\mathcal{A}_1$ for solving A_1 which uses as certain elementary steps an algorithm $\mathcal{A}_2$ which solves A_2. Moreover, if $A_2 \in \mathcal{P}$, then $A_1 \in \mathcal{P}$ also.

A special kind of polynomial reduction is of interest.

We say that a recognition problem A_1 can be polynomially transformed to a recognition problem A_2 if and only if an input x (of arbitrary dimension) of A_1 can be transformed in polynomial time (relative to the input dimension of x) into an input y of A_2 such that if x is an input of A_1 with answer 'yes', then y is an input of A_2 with answer 'yes', while if x gives answer 'no', then y gives answer 'no'. It is clear from this definition that polynomial transformation of problems is a particular case of polynomial reduction.

The study of combinatorial problems in $\mathcal{NP}$ using polynomial reducibility has made it possible to distinguish a subset of problems in $\mathcal{NP}$ which have the property that they have the same complexity.

An recognition problem $A \in \mathcal{NP}$ is said to be an $\mathcal{NP}$-*complete recognition problem* if all other problems in $\mathcal{NP}$ can be polynomially transformed to A.

The class of $\mathcal{NP}$-complete problems can be characterized by the following important property:

> If there is a polynomial algorithm for solving at least one $\mathcal{NP}$-complete combinatorial problem, then there are polynomial algorithms for solving all $\mathcal{NP}$-complete problems.

Up till now no $\mathcal{NP}$-complete problem can be solved by any polynomial algorithm known. Taking this into account, many experts adhere to the conjecture that there cannot exists a polynomial solution algorithm for any $\mathcal{NP}$-complete problem. A proof of this conjecture has not yet been given.

To prove that a combinatorial problem belongs to the class of $\mathcal{NP}$-complete problems it suffices:

- to prove that it belongs to $\mathcal{NP}$;
- to prove that some $\mathcal{NP}$-complete problem can be reduced to this problem.

The first problem which was proved to belong to the class of $\mathcal{NP}$-complete problems is the problem of 'realizability'. Its formulation and the proof of its belonging to the class of $\mathcal{NP}$-complete problems is given in [21]. The combinatorial 'realizability' problem can be formulated for Boolean variables, and is one of the six fundamental $\mathcal{NP}$-complete problems which are commonly used for proving the $\mathcal{NP}$-completeness of new combinatorial problems. In this list occur five other problems; these are given in Examples 1.9, 1.13, 1.16, 1.18, and 1.21. Methods of polynomial reduction and proofs of $\mathcal{NP}$-completeness, as well as the set of $\mathcal{NP}$-complete combinatorial problems in full extent are given in [50].

The practical value of our theory concerning the classification by complexity of combinatorial problems lies in the fact that when a new combinatorial problem emerges, it is possible to estimate its relative complexity. This, in turn, allows us to introduce a certain clarity in the choice of methods for solving the problem stated.

CHAPTER 2

Extremal problems on packability of number partitions

This Chapter contains the basic mathematical results of the investigation on packability of number partitions and presents the most complete list of results in this direction known at present. As an application of the applicability of these results we outline their relations with an archaic question on weighings and other problems.

1. Number partitions

In practice one often has to solve problems which involve natural numbers and sums of such. A convenient combinatorial interpretation of such problems is the notion of number partition. Number partitions, as an individual mathematical notion, first appeared in correspondence between J. Bernoulli and G. Leibniz. At the time of their appearance, partitions were the traditional object of enumeration problems in combinatorics, and served as a powerful stimulus for the development of its methods, in the first instance of its enumeration methods. Quite recently one has succeeded in extending their domain of use to extremal problems.

Apparently, one of the earliest proper extremal number-theoretical results is due to J. Sylvester, whose theorem asserts: Let $r_1, \ldots, r_t$ be mutually prime prime natural numbers, and let $s(r_1, \ldots, r_t)$ be the largest integer s that cannot be represented in the form

$$s = \sum_{i=1}^{t} a_i r_i, \qquad \text{where } a_i \in \mathbf{N}_0 = \{0, 1, 2, \ldots\}, \; i = 1, \ldots, t.$$

Then

$$s(r_1, r_2) = r_1 r_2 - r_1 - r_2.$$

For $t \geq 3$ the problem of computing the exact value of $s(r_1, \ldots, r_t)$ is still open, and is called the *Frobenius problem*.

To give a feeling for this problem we turn to a particular result:

$$s(n, n+1, n+2) = n \left[\frac{n}{2} \right] - 1.$$

It is proved as follows. We denote the righthand side of this equation by $S(n)$, and prove that it cannot be represented as a function of $n, n+1, n+2$. Assume the contrary:

$$S(n) = an + b(n+1) + c(n+2) = (a+b+c)n + (b+2c).$$

Then $a + b + c \leq \left[\frac{n}{2}\right] - 1$ and $b + 2c \geq n - 1$, so that $c \geq \left]\frac{n}{2}\right[$, contradicting the representation $S(n) = n\left[\frac{n}{2}\right] - 1 < \left]\frac{n}{2}\right[(n+2)$.

We now verify that $S(n)$ is the largest number that cannot be represented such, i.e. that every number larger than it is representable in terms of $n, n+1, n+2$. Each natural number larger than $S(n)$ can be written as

$$n\left(\left[\frac{n}{2}\right] + x\right) - i, \qquad x = 1, 2, \ldots; \quad i = 1, \ldots, n.$$

We now consider four cases:

n odd, $i = 1, \ldots, \left[\frac{n}{2}\right] + 1$; then

$$n\left(\left[\frac{n}{2}\right] + x\right) - i = \left(\frac{n-1}{2} - i - 1\right)(n+2) + (i-1)(n+1) + (x-1)n.$$

n odd, $i = \left[\frac{n}{2}\right] + 2, \ldots, n$; then

$$n\left(\left[\frac{n}{2}\right] + x\right) - i = (n-i)(n+1) + \left(i + x - \left[\frac{n}{2}\right]\right)n.$$

n even, $i = 1, \ldots, \frac{n}{2}$; then

$$n\left(\left[\frac{n}{2}\right] + x\right) - i = \left(\frac{n}{2} - i\right)(n+2) + i(n+1) + (x-1)n.$$

n odd, $i = \frac{n}{2}, \ldots, n$; then

$$n\left(\left[\frac{n}{2}\right] + x\right) - i = (n-i)(n+1) + \left(i + x - \frac{n}{2}\right)n.$$

This proves the required. This proof was obtained together with A. Klimov and I. Kan as a result of the analysis of data of a special package of computer programs, created for the computation of exact formulas in the Frobenius problem. For not simply numerical values of s but precise formulas, see Problem 2.26.

Thus, even a question of representing one number by a linear combination of several numbers turns out to be nontrivial. What happens to the presentation of sums by other sums?

1.1. Basic concepts and definitions. A *partition* of a natural number n is a representation of it as an unordered sum of natural numbers: $n = n_1 + \cdots + n_r$; these numbers n_i are called the *parts of the partition*, and their number r is called the *rank of the partition*.

A *composition* is a representation of a natural number n as an ordered sum of natural numbers. Thus, a composition may be regarded as an 'ordered partition'.

For example, for $n = 6$ the partitions are:

> **rank one:** $6 = 6$;
> **rank two:** $6 = 5 + 1$, $6 = 4 + 2$, $6 = 3 + 3$;
> **rank three:** $6 = 4 + 1 + 1$, $6 = 3 + 2 + 1$, $6 = 2 + 2 + 2$;
> **rank four:** $6 = 3 + 1 + 1 + 1$, $6 = 2 + 2 + 1 + 1$;
> **rank five:** $6 = 2 + 1 + 1 + 1 + 1$;
> **rank six:** $6 = 1 + 1 + 1 + 1 + 1 + 1$.

The compositions of $n = 6$ are:

rank one: $6 = 6$;

rank two: $6 = 5 + 1$, $6 = 4 + 2$, $6 = 3 + 3$,
$6 = 1 + 5$, $6 = 2 + 4$;

rank three: $6 = 4 + 1 + 1$, $6 = 3 + 2 + 1$, $6 = 2 + 2 + 2$,
$6 = 1 + 4 + 1$, $6 = 3 + 1 + 2$,
$6 = 1 + 1 + 4$, $6 = 2 + 3 + 1$,
$6 = 2 + 1 + 3$,
$6 = 1 + 3 + 2$,
$6 = 1 + 2 + 3$;

rank four: $6 = 3 + 1 + 1 + 1$, $6 = 2 + 2 + 1 + 1$,
$6 = 1 + 3 + 1 + 1$, $6 = 2 + 1 + 2 + 1$,
$6 = 1 + 1 + 3 + 1$, $6 = 2 + 1 + 1 + 2$,
$6 = 1 + 1 + 1 + 3$, $6 = 1 + 2 + 2 + 1$,
$6 = 1 + 2 + 1 + 2$,
$6 = 1 + 1 + 2 + 2$;

rank five: $6 = 2 + 1 + 1 + 1 + 1$,
$6 = 1 + 2 + 1 + 1 + 1$,
$6 = 1 + 1 + 2 + 1 + 1$,
$6 = 1 + 1 + 1 + 2 + 1$,
$6 = 1 + 1 + 1 + 1 + 2$;

rank six: $6 = 1 + 1 + 1 + 1 + 1 + 1$.

It is easy to find the number of rank-r compositions of a natural number n: it is the number of ways of distributing $r - 1$ strokes in the $n - 1$ intervals between n points. It is $\binom{n-1}{r-1}$. If the rank is not fixed, then a stroke may be distributed in each of the intervals or not, and the total number of compositions of a number n is thus 2^{n-1}. Consequently, compositions and partitions may be regarded as ordered and unordered multisets, respectively, with as elements natural numbers. In correspondence with this, a partition is often written in vector notation, $(n_1, \ldots, n_r) \vdash n$, which denotes that $n = n_1 + \cdots + n_r$, or in shortened notation $(n_1^{a_1}, \ldots, n_r^{a_r}) \vdash n$, meaning that the part n_i occurs in this partition a_i times, i.e. $n = a_1 n_1 + \cdots + a_r n_r$, and the rank of this partition is $a_1 + \cdots + a_r$. Thus, every partition can be represented as $(1^{m_1}, 2^{m_2}, \ldots, n^{m_n}) \vdash n$, where m_i is a nonnegative integer, indicating how often the number i occurs in this partition of n, i.e. $n = \sum_{i=1}^{n} i m_i$, and $\sum_{i=1}^{n} m_i$ is the rank of this partition. We may depict a partition graphically by a point diagram, called *Ferrer diagram*, e.g.

$$(1, 2^2, 3^2, 4) \Leftrightarrow \begin{matrix} \cdot \\ \cdot\, \cdot \\ \cdot\, \cdot \\ \cdot\, \cdot\, \cdot \\ \cdot\, \cdot\, \cdot \\ \cdot\, \cdot\, \cdot\, \cdot \end{matrix}$$

This picture is convenient for representing various transformations of partitions.

For example, if the given diagram is rotated, we obtain a Ferrer diagram of the form

$$\begin{array}{l} \cdot \\ \cdot\,\cdot\,\cdot \\ \cdot\,\cdot\,\cdot\,\cdot\,\cdot \\ \cdot\,\cdot\,\cdot\,\cdot\,\cdot\,\cdot \end{array}$$

which clearly corresponds to the partition $(1, 3, 5, 6)$, called the *conjugate partition* of $(1, 2^2, 3^2, 4)$.

Partition problems are more difficult than the corresponding composition problems. For example, even the computation of $p(n, r)$, the number of rank-r partitions of n, i.e. the calculation of the number of solutions of the equation $n = x_1 + \cdots + x_r$, $x_1 \geq \cdots \geq x_r$, in natural numbers x_i is one of the fundamental moments of enumeration partition theory, and one may get acquainted with it through the classical handbooks of MacMahon and Andrews. Our main problem, however, is to get acquainted with the theory of partitions as an object in extremal combinatorial problems.

Next to conjugation there are other kinds of correspondences between partitions. A sufficiently general form of extremal problem concerning partitions can be stated as follows: How may partitions can there be in a given correspondence? The selection of a concrete correspondence between partitions is determined by the conditions of the practical problem; more precisely, we should select a correspondence for which the partitions serve as combinatorial scheme for the problem. Therefore, below we introduce and study correspondences between partitions by means of which a wide circle of important applied problems can be solved.

For a long time, extremal problems on number partitions did not form a separate direction, although the main underlying facts appeared much earlier. It is useful to turn our attention to one of them; more precisely, to a particular instance of a very general theorem which is now called in honor of its author, the English logician Ramsey. To this end we define a correspondence $\succeq$ between partitions by the rule: a partition $(n_1, \ldots, n_r)$ is in correspondence $\succeq$ with a partition $(k_1, \ldots, k_r)$ if there is an i $(1 \leq i \leq r)$ such that $k_i \leq n_i$, i.e. if the second partition has a part not exceeding the corresponding part of the first partition. We can then ask: For a given partition $(k_1, \ldots, k_r)$, what is the smallest $n = n(k_1, \ldots, k_r)$ for which for any partition of this n into r parts the correspondence $(n_1, \ldots, n_r) \succeq (k_1, \ldots, k_r)$ will hold? It is not difficult to verify that this smallest $n(k_1, \ldots, k_r)$ exists, and can be computed by the formula

$$n(k_1, \ldots, k_r) = \sum_{i=1}^{r} k_i - r + 1. \tag{2.1'}$$

We can also compute the converse characteristic: For a given partition $(n_1, \ldots, n_r)$, compute the largest $k = k(n_1, \ldots, n_r)$ each partition of which into r parts will have the property that $(n_1, \ldots, n_r) \succeq (k_1, \ldots, k_r)$. This characteristic can be computed by

$$k(n_1, \ldots, n_r) = \sum_{i=1}^{r} n_i + r - 1. \tag{2.2'}$$

We prove both formulas. The quantity $n = n(k_1, \ldots, k_r)$ cannot be smaller than the righthand side of (2.1'), since in that case there would be a partition for which the required correspondence would not hold:

$$\sum_{i=1}^{r} k_i - r \dashv (k_1 - 1, \ldots, k_r - 1) \preceq (k_1, \ldots, k_r).$$

If now n would equal the righthand side of (2.1'), then in any partition $(n_1, \ldots, n_r) \vdash n$ we can find a part $n_i \geq k_i$, contradicting that $n_i \leq k_i - 1$ $(i = 1, \ldots, r)$, and thus we obtain the contradictory system of inequalities

$$\sum_{i=1}^{r} k_i - r = \sum_{i=1}^{r}(k_i - 1) \geq n = \sum_{i=1}^{r} k_i - r + 1.$$

The quantity $k = k(n_1, \ldots, n_r)$ cannot be larger than the righthand side of (2.2'), since in this case we could find a partition for which the required correspondence would not hold:

$$(n_1, \ldots, n_r) \succeq (n_1 + 1, \ldots, n_r + 1) \vdash \sum_{i=1}^{r} n_i + r.$$

If now k equals the righthand side of (2.2'), then $n = k - r + 1$, and hence, by (2.1'), the required correspondence with all rank-r partitions of k is satisfied by all rank-r partitions of n, and not only by $(n_1, \ldots, n_r)$. This means that if $k = k(n_1, \ldots, n_r)$ is expressed as in (2.1'), then every rank-r partition of k is in correspondence with every rank-r partition of n. In particular, (2.1') immediately implies the well-known

Dirichlet principle: If $n(k, r)$ is the smallest integer n such that in each rank-r partition of n there is apart not less than k, then

$$n(k, r) = rk - k + 1. \tag{2.3'}$$

In fact, it suffices to put $k_i = k$ $(i = 1, \ldots, r)$ in (2.1').

This principle is often stated in terms of allocations, and is then called the

Principle of boxes: For any allocation of $r + 1$ objects over r boxes, there is a box containing at least two objects.

In fact, it suffices to put $k = 2$ in (2.3'), and to note that each allocation of n indistinguishable objects over r indistinguishable boxes can be adequately represented by a partition of n into at most r parts. Thus, (2.1') generalizes Dirichlet's principle, but already cannot be treated in terms of allocations. However, there is a correspondence between partitions that can be treated well in terms of allocations.

1.2. Four questions. Here we give four concrete problem statements, which in the course of the book will be called upon to aid in the perception of the material.

QUESTION 2.1. How many weights does one minimally need to weigh any integer number of pounds from 1 to k?

On equi-arm beam balances we provide for two kinds of precise weighings—single pan and double pan. In the first case the weights may be put into one pan only; in the second they may be put into both.

In his book '*Analysis Infinita*', more precisely, in the chapter on number partitions, L. Euler used the method of generating functions to substantiate the efficiency of two well-known fastest growing sequences of weights $\{(p+1)^i\}_{i=0,1,2,\ldots}$ $(p = 1, 2)$ for p-pan weighing, respectively. Of course, other authors too have singled out precisely these sequences, since these are the most efficient systems of weights for weighing any integer weight. In the finite case (load not exceeding k) it is natural to assume that the total weight of the weights is k; in this case geometric progressions are far from always being efficient.

QUESTION 2.2. How many edges, $m(n, H_k)$, must an n-vertex graph G_n minimally have, if among any k vertices there is a subgraph isomorphic to a prescribed k-vertex graph H_k?

For example, imagine that one has a board with n terminals and that it is required to join these terminals by wires such that any k terminals are mutually 'connected'. In other words, representing the terminals by vertices and the wires by edges, we arrive at the necessity of constructing a graph in which every k vertices are joined by at least one cycle. Here, of course, it is natural to minimize the total number of wires.

QUESTION 2.3. What is the least amount of objects, of any kind, needed to realize all outcomes in the scheme of allocating n indistinguishable particles over r indistinguishable boxes?

At first glance this seems paradoxical; the problem is best understood by considering a concrete numerical example. Let $n = 6$ and $r = 2$. Then all possible outcomes of this allocation scheme are: $(6,0)$, $(5,1)$, $(4,2)$, $(3,3)$, where each number indicates the amount of particles in each box. If we now consider three groups of 6 particles with 1, 2, and 3 particles in each of them, respectively, then a direct verification shows that the outcomes of the initial scheme may be realized not only by the 6 particles, but already by these three groups: e.g., $(4,2) = (3+1,2)$, $(3,3) = (2+1,3)$, etc. Therefore the question concerning the least possible number of such groups makes sense.

QUESTION 2.4. In the process of functioning, the memory of a computer becomes 'scattered' into occupied and free sectors—so-called fragments. If we now have to feed new information into the computer's memory, e.g. programs or data arrays, requiring memory sizes $k_1, \ldots, k_t$, then there arises the natural question: Can these sizes be distributed over the free memory fragments of sizes $n_1, \ldots, n_r$?

Computer memory fragmentation is not just a simple concrete situation requiring solution for a given selection of sizes of fragments and inquiries, it is a phenomenon lying at the basis of a number of important processes in the actual functioning of the computer's memory. Thus, it is also at the basis of developing theoretical approaches to the investigation of this phenomenon. The process of dynamic computer memory allocation is an example of most general, but not simple, fragmentation (see Problem 2.21).

1.3. Packability of partitions. The main correspondence between number partitions which we will investigate is packability of them.

A *partition* $(k_1, \ldots, k_t)$ *can be packed into a partition* $(n_1, \ldots, n_r)$ if there is a map $\phi \colon \{1, \ldots, t\} \to \{1, \ldots, r\}$ such that the following system of inequalities holds:

$$\sum_{j \in \phi^{-1}(i)} k_j \leq n_i, \qquad i = 1, \ldots, r, \tag{2.4'}$$

where $\phi^{-1}(i) = \{j \colon j \in \{1, \ldots, t\}, \phi(j) = i\}$ is the complete pre-image of the element i under ϕ.

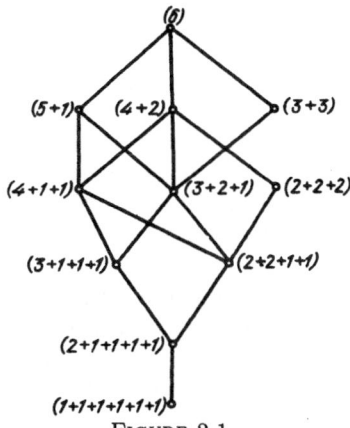
FIGURE 2.1

In other words, the partition $(k_1, \ldots, k_t)$ can be packed into the partition $(n_1, \ldots, n_r)$ if the parts k_i of $(k_1, \ldots, k_t)$ can be grouped together into r groups (each part k_i goes into one group, and empty groups are omitted) in such a way that after adding all parts k_i in each group we obtain r numbers $p_i \leq n_i$ $(i = 1, \ldots, r)$. If the partition $(k_1, \ldots, k_t)$ can be packed into the partition $(n_1, \ldots, n_r)$, then we write, using the set inclusion sign, $(k_1, \ldots, k_t) \subseteq (n_1, \ldots, n_r)$. For example, $(2, 2, 2) \subseteq (4, 2)$, since $(4, 2) = (2 + 2, 2)$; however, $(2, 2, 2) \not\subseteq (3, 3)$, since the three pairs cannot be grouped into two groups each of which does not exceed three.

Packability is a binary relation on the set of all partitions of all natural numbers. It is not difficult to verify that this binary packability relation has the following properties:

a) reflexivity: $(n_1, \ldots, n_r) \subseteq (n_1, \ldots, n_r)$;

b) antisymmetry: if $(k_1, \ldots, k_t) \subseteq (n_1, \ldots, n_r)$, $(n_1, \ldots, n_r) \subseteq (k_1, \ldots, k_t)$, then $(k_1, \ldots, k_t) = (n_1, \ldots, n_r)$;

c) transitivity: if $(k_1, \ldots, k_t) \subseteq (m_1, \ldots, m_l)$, $(m_1, \ldots, m_l) \subseteq (n_1, \ldots, n_r)$, then $(k_1, \ldots, k_t) \subseteq (n_1, \ldots, n_r)$.

Consequently, packability is a partial order relation on the set of number partitions.

We introduce the following notation:

P: the set of all partitions of all natural numbers;

$P(n)$: the set of all partitions of the number n;

P_r: the set of all partitions of rank r;

$P_r(n)$: the set of all partitions of rank r of the number n.

Consequently, $P_r(n) = P(n) \cap P_r$.

We regard the sets P, P_r, and $P(n)$ as being ordered by packability; the set $P_r(n)$ can be conveniently regarded as being lexicographically ordered.

The Hasse diagram of $P(6)$ is depicted in Figure 2.1. From it we can clearly see that the partially ordered set $P(n)$ has a largest and a smallest element: these are (6) and (1^6), respectively. The levels in the Hasse diagram are the sets $P_r(n)$, which are depicted in lexicographical order.

In essence, the main question concerning packability of number partitions consists of establishing the packability of a fixed partition $(k_1, \ldots, k_t)$ into another fixed partition $(n_1, \ldots, n_r)$. In other words, is there a packing

$$(k_1, \ldots, k_t) \subseteq (n_1, \ldots, n_r)?$$

Next to this question of testing the possibility of packing arises the question of realizing it: how fast can the packing of one partition into another be realized? It is obvious that these questions are algorithmically equivalent, since the presence of a fast testing algorithm ensures the presence of a corresponding packing algorithm, and conversely.

We estimate the complexity of a complete listing in order to determine packability of two concrete partitions. By the definition of packability, this listing reduces to the listing of all maps $\phi \colon \{1, \ldots, t\} \to \{1, \ldots, r\}$ and the verification of the system of r inequalities (2.4') for each of these maps. Since the complete pre-image of each such map ϕ is an ordered list of t not necessarily distinct elements, taking any of the r values, by the listing scheme the total number of such maps is r^t. Consequently, a complete listing in order to determine packability consists of the verification of r^t systems of inequalities (2.4'), or of the verification of the r^{t+1} inequalities making up the system (2.4').

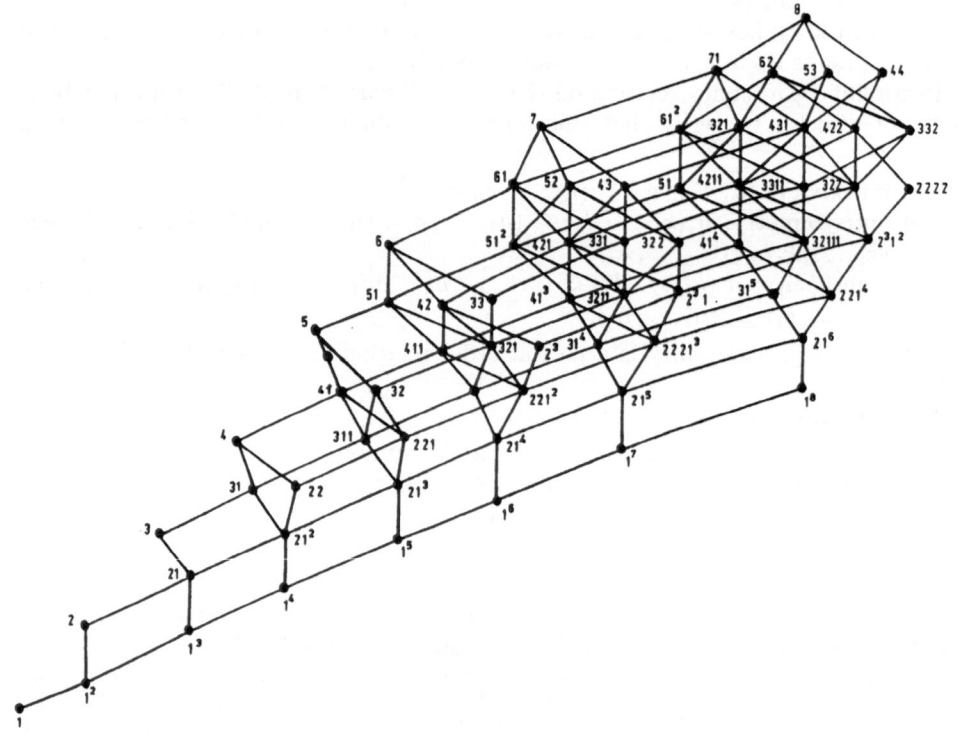

FIGURE 2.2

2. Hasse diagrams and Möbius packability functions

It turns out to be useful to draw the Hasse diagram of the whole set P and some of its subsets. (See Figure 2.2.) It is clear from this drawing how to construct the Hasse diagram of P: first one draws the two independent diagrams of the sets $P(n)$ and $P(n+1)$, with the first one level lower than the second, and then one draws the edges $\{(n_1, \ldots, n_r), (n_1, \ldots, n_r, 1)\}$ between them, and only these edges. In fact, if $q \in P(n)$, $p, (q, 1) \in P(n+1)$, and $p \notin U^+(q, 1) = \{x \in P : (q, 1) \subseteq x\}$ and $q \subseteq p$, then for any concrete packability of q into p only one unit volume remains unfilled, which means that $(q, 1) \subseteq p$, contradicting the assumption. Thus, the Hasse diagram of P can be represented by the sequence of Hasse diagrams of the sets $P(n)$, $n = 1, 2, \ldots$, and edges between them as indicated. These edges can be naturally interpreted as a 'shift' of $P(n)$ inside $P(n+1)$ together with a raise of one level. It turns out that for an arbitrary interval this procedure gives information on the Möbius function also.

LEMMA 2.1 (SHIFT LEMMA). *Suppose the interval $[x', y']$ is the result of 'shifting' an interval $[x, y]$. Then*

$$\mu(x, y') = -\mu(x, y).$$

PROOF. We proceed by induction with respect to $n = |[x, y]|$. If $n = 1$, then $[x, y']$ is a chain of length 1. Consider an arbitrary $n > 1$. Then

$$\mu(x, y') = -\sum_{x \le z < y'} \mu(x, z) = -\left(\sum_{x \le z \le y} \mu(x, z) + \sum_{\substack{z < y' \\ z \in [x', y']}} \mu(x, z) \right) =$$

$$= -\sum_{\substack{z < y' \\ z \in [x', y']}} \mu(x, z) = \text{ind} = -\sum_{\substack{z < y' \\ z \in [x, y]}} -\mu(x, z) = \sum_{x \le z < y} \mu(x, z) = -\mu(x, y). \qquad \square$$

In particular, this simple, but very general, result implies that all information on the Möbius function of P is concentrated on the subsets $P(n)$ only. In addition we need only use the following lemma.

In fact, this lemma is an instance of the theorem on the Möbius function of the discrete product of two partially ordered sets, given in the book of R. Stanley.

LEMMA 2.2 (LEMMA ON THE ZERO ZONE). *If $(P, \le)$ is a locally finite partially ordered set, and if $D \subset (P, \le)$ is a subset such that for a fixed element $p \in (P, \le)$,*

$$\forall q \in D \exists q_1 \in [p, q] : [p, q] \setminus [p, q_1] \subset D,$$

then

$$\forall q \in D : \mu(p, q) = 0.$$

PROOF. Assume the contrary and consider the smallest $q \in D$ satisfying the requirement but for which $\mu(p, q) \ne 0$. Then

$$0 \ne \mu(p, q) = -\sum_{p \le z < q} \mu(p, z) = -\left(\sum_{z \in [p, q_1]} \mu(p, z) + \sum_{\substack{z < q \\ z \in [p, q] \setminus [p, q_1]}} \mu(p, z) \right).$$

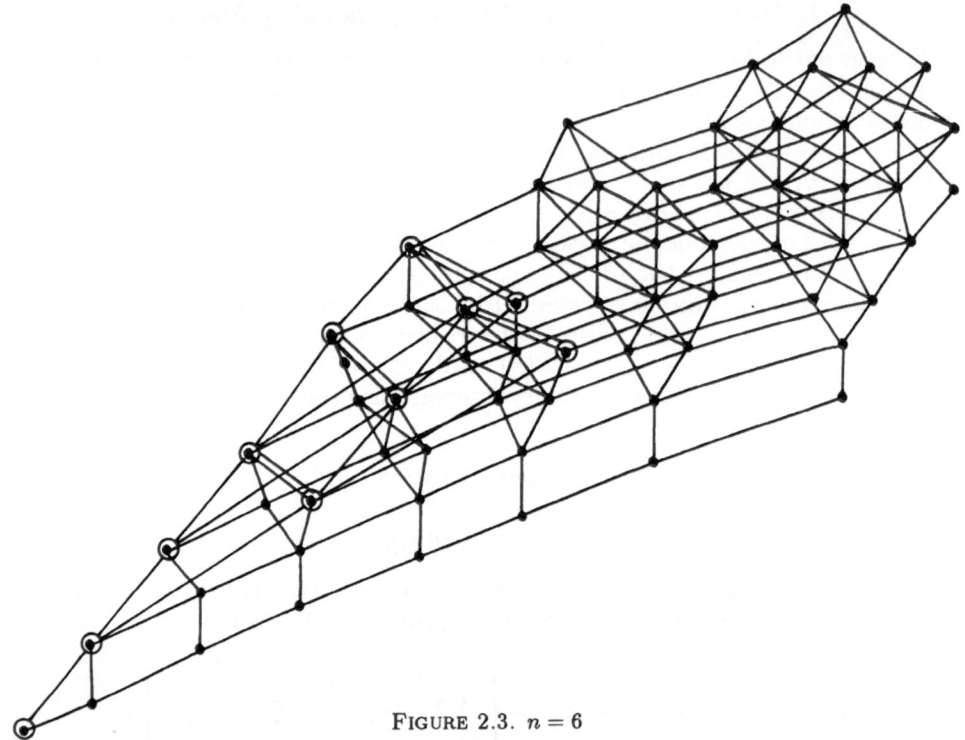

FIGURE 2.3. $n = 6$

Here q_1 is precisely an element for which the condition mentioned in the requirements is fulfilled. Therefore the last sum must vanish: either the summation range is empty or all terms are equal to zero, since the range of summation lies entirely inside D and consists of elements of levels less than q, by the choice of q. Since the one but last sum is identically zero by the definition of the Möbius function, the total sum must equal zero, contradicting the assumption. $\square$

It is easily seen from the diagram how the levels of the diagram of P are made up:

$$W_t(P) = \bigcup_{l \geq 0} P_{t-2l}(t - l), \qquad t = 1, 2, \ldots.$$

Altogether, the high degree of visuality of this diagram of P is clear from the following example. Consider $P(\leq n)$, the set of all partitions of all numbers not larger than n, and consider in it all partitions without unit parts, as well as the partition (1). It is clear from the diagram that the thus distinguished subset is isomorphic to the set $P(n)$. This follows in particular from the well-known identity

$$|P(n)| - 1 = \sum_{k=2}^{n} |P(k; \geq 2)|,$$

where $P(k; \geq 2)$ is the set of all partitions of the number k without unit parts.

For $n = 6$ see Figure 2.3.

3. Simplest properties of packability of number partitions

Next to packability one has also studied other binary relations on partitions. Some particular cases of packability have been considered earlier. For example, on $P(n)$ ordering by packability has also been called domination by either decomposition or pasting. In these particular instances one studied, as a rule, either enumeration or structural problems related to partitions (see, e.g., [127]). But the generality of $P(n)$ is not sufficient for a convenient statement of extremal problems. In the given generality the notion of packability was introduced in [5] in connection with the modeling of a number of engineering phenomena.

The question of determining packability for two concrete partitions is an algorithmically hard problem, since this problem reduces to the solution of a system of diophantine equations [5]; also, it is equivalent to one of the problems on the list of algorithmically hard problems (the so-called *bin-packing problem* [21]). However, sometimes one succeeds in reducing the number of input parameters.

3.1. The exchange lemma. Each number partition may be interpreted as a multiset; more precisely, to a partition $(1^a, 2^b, 3^c, \dots)$ corresponds the multiset $\{1^a, 2^b, 3^c, \dots\}$, and conversely. This makes it possible to use the operations of union, intersection, and difference on partitions. For example, if p and q are two partitions of a single number, then $p \cap q$ denotes the partition obtained as the result of intersecting the multiset of all parts of p with that of q. The use of these operations in the study of packability of partitions makes it possible to detect a number of most simple properties of packability.

LEMMA 2.3. *A partition q can be packed into a partition p if and only if the partition $q - (p \cap q)$ can be packed into the partition $p - (p \cap q)$.*

In other words, the question of packability of partitions is equivalent to the question of the packability of the partitions obtained from the original ones by deleting in each of them an identical amount of identical parts. For example, if $(1, 2, 2, 3, 5) \subset (2, 3, 3, 7)$, then $(1, 2, 5) \subset (3, 7)$, and conversely.

PROOF. Sufficiency is obvious, so we prove necessity. Suppose the partition $q = (q_1, \dots, q_r)$ can be packed into the partition $p = (p_1, \dots, p_t)$, i.e. the system of inequalities

$$p_i \geq q_{i,1} + \cdots + q_{1,l} \tag{2.1}$$

holds, where $1 \leq i \leq t$ and $q = (q_{1,1}, q_{1,2}, \dots, q_{1,l}, q_{2,1}, \dots, q_{t,l})$.

If in (2.1) we have equality for at least one i, then we can immediately pass to partitions with fewer parts. If now for some q_i there is a p_j such that $p_j = q_{i,m}$, then the jth inequality can be taken out from (2.1), and the ith inequality can be written as

$$p_i \geq q_{i,1} + \cdots + q_{i,m} + \cdots + q_{i,l} = q_{i,1} + \cdots + p_j + \cdots + q_{i,l} \geq$$
$$\geq q_{i,1} + \cdots + q_{i,m-1} + q_{j,1} + \cdots + q_{j,l} + q_{i,m+1} + \cdots + q_{i,l}.$$

Thus, we pass to a partition with lower rank. Since this process is finite, we obtain the required. $\square$

It is obvious that the practical value of the exchange lemma is restricted by the complexity of distinguishing an identical amount of identical parts in two fixed partitions.

However, one way or the other, the practical need to solve the question of packability of partitions has not completely disappeared. Moreover, a number of actual situations do not leave the time to perform a complete listing. It is thus necessary to search for faster methods of verifying a guaranteed packability. For example, if we are to verify packability of a partition $(k_1, \ldots, k_t) \vdash k$ into a partition $(n_1, \ldots, n_r) \vdash n$ and it is known that $k > n$, then an answer is obvious even without complete listing: the first partition can not be packed into the second. It is precisely this trivial reason that lies at the basis of the here presented extremal approach to constructing fast methods for verifying a guaranteed packing. In this sense the verification of the inequality $k > n$ is no other than the verification of an extremal (in fact, largest possible) number k, since n is the largest possible value of k for which there can be a partition of k that can be packed into a partition of n. Thus, the extremal approach to establishing packability consists in detecting and computing the extremal characteristics of partitions (which are nontrivial, but have polynomial complexity) whose comparison gives a guaranteed condition for packability or nonpackability.

The first extremal result concerning packability of partitions is:

3.2. The rank condition for packability. In a number of cases we succeed in solving immediately, without any algorithmic verification, the combinatorial problem of recognizing packability of partitions. For example, the following theorem saves us from the necessity of clarifying the question of packability of partitions of certain fixed ranks.

THEOREM 2.1. *Let $t(n, k, r)$ be the smallest number t for which*

$$\forall p \in P_r(n) \ \forall q \in P_t(k) : \ q \subset p. \tag{2.2}$$

Then

$$t(n, k, r) = \max\left(k - \left]\frac{n}{r}\right[+ 1, 1\right). \tag{2.3}$$

PROOF. If $\max(k -]n/r[+ 1, 1) = 1$, then $k \le]n/r[$ and by Dirichlet's principle there is in any partition of n into r parts a part not less than k, i.e. the required packability holds.

Suppose now that

$$\max\left(k - \left]\frac{n}{r}\right[+ 1, 1\right) = k - \left]\frac{n}{r}\right[+ 1.$$

Then the truth of the packing

$$(k - t + 1, 1, \ldots, 1) \subset \left(\left]\frac{n}{r}\right[, \ldots, \left]\frac{n}{r}\right[\right) \in P_r(n)$$

implies the truth of the inequality $t \ge k -]n/r[+ 1$. We prove this case by induction with respect to n.

Assume that (2.3) holds for all values up to $n-1$ inclusive. We have to prove it for n. Let $t = k -]n/r[+ 1$, $p = (m, p_{r-1}(n - m)) \in P_r(n)$, $q = (d, p_{t-1}(k - d)) \in P_t(k)$.

Clearly, $m \geq]n/r[= k - t + 1 \geq d$. If $m = d = k - t + 1$, then $q = (k - t + 1, 1^{t-1})$, and thus $\forall p \in P_r(n)$, $p \subset q$, so that $q = (1^t)$. So we assume $m > d > 1$.

The packing $q \subset p$ follows from the packing $p_{t-1}(k - d) \subset (m - d, p_{r-1}(n - m))$, which, in turn, follows from the inequality $t - 1 \geq t(n - d, k - d, r)$. By the induction hypothesis, this inequality takes the form $k -]n/r[= t - 1 \geq t(n - d, k - d, r) = k - d -](n - d)/r[+ 1$, or $]n/r[+ 1 \leq d +](n - d)/r[$, which holds for $d \geq 2$. $\square$

It is obvious how to use this theorem in establishing packability for two partitions. If we are to verify the packability of a partition $(k_1, \ldots, k_t)$ into a partition $(n_1, \ldots, n_r)$, then fulfillment of the inequality $t \geq t(n, k, r)$ implies $(k_1, \ldots, k_t) \subseteq (n_1, \ldots, n_r)$.

COROLLARY 2.1. *If $n(k, t, r)$ is the smallest n for which*

$$\forall p \in P_r(n) \ \forall q \in P_t(k) : \ q \subset p,$$

then

$$n(k, t, r) = \max(k, r(k - t) + 1).$$

The next step in answering the question of packability of two concrete partitions lies in dropping the freedom of choosing the partition to be packed. In other words, in fixing the 'smaller' partition by choosing from the partitions into which packing is to be realized. Thus, the answer to the corresponding extremal problem will in this case depend not on three parameters, as in Corollary 2.1, but on $t + 1$ parameters (all parts of the partition $(k_1, \ldots, k_t) \vdash k$ to be packed and the rank r of the partitions in which this extremal bound must guarantee the packing $(k_1, \ldots, k_t) \vdash k$).

3.3. Sal'nikov constants. An essential widening of the class of extremal problems regarding number partitions is obtained by the results and problem statements of S.G. Sal'nikov. Below we give an exposition of his results and statements.

Let $P(k, p) = \{(1^{\alpha_1}, \ldots, p^{\alpha_p}) \vdash k\}$ be the set of all partitions with parts at most p. Clearly, $P(k, 1) = \{(1, \ldots, 1)\}$, $P(k, k) = P(k)$, and, in general, $P(k, p + 1) \supset P(k, p)$.

The structure of $P(k, p)$ is described in the following assertion:

> if $\bar{k}'$ and $\bar{k}''$ are two partitions of a number k, the partition $\bar{k}'$ is lexicographically smaller than $\bar{k}''$, and $\bar{k}''$ belongs to $P(k, p)$, then $\bar{k}'$ also belongs to $P(k, p)$.

Let k, p, r be natural numbers. The constant $n(k, p, r)$ is defined as the smallest n such that $\forall \bar{k} \in P(k, p) \ \forall \bar{n} \in P_r(n) : \ \bar{k} \subset \bar{n}$.

Prove that, for natural numbers k, p, r $(1 \leq p \leq k)$,

$$n(k, p, r) = \max \left(p \left\lfloor \frac{k}{p} \right\rfloor + (p - 1)(r - 1), \ k + \left(\left\lfloor \frac{k}{\lceil k/p \rceil} \right\rfloor - 1 \right) (r - 1) \right).$$

Let $r(k, p, n) = \max\{r : \forall \bar{k} \in P(k, p) \ \forall \bar{n} \in P_r(n): \ \bar{k} \subset \bar{n}\}$. The following equation readily follows from the definitions of $n(k, p, r)$ and $r(k, p, n)$:

$$r(k, p, n) = \max\{r : \ n(k, p, r) \leq n\}.$$

Using the formula for $n(k, p, r)$ and this equality, calculate an explicit expression for $r(k, p, n)$.

For an arbitrary (fixed) partition

$$(k_1, \ldots, k_t) \vdash k,$$

the function $p = p(k_1, \ldots, k_t)$ is defined as the largest integer p having the property that for any partition $\overline{k}' = (1^{\alpha_1}, \ldots, p^{\alpha_p}) \vdash k$ the packing $\overline{k}' \subset (k_1, \ldots, k_t)$ holds.

Prove that the 'order' equality

$$P(k) \asymp \ln(k)$$

holds, i.e. there are real constants $c'' > c' > 0$ (independent of k) such that

$$c' \ln(k) < P(k) < c'' \ln(k).$$

4. The principle of complete packability

THEOREM 2.2 (PRINCIPLE OF COMPLETE PACKABILITY). *Let* $k_1 \geq k_2 \geq \cdots \geq k_t$ *and* r *be natural numbers, and let* $n(k_1, \ldots, k_t; r)$ *be the smallest* n *for which the partition* $(k_1, \ldots, k_t) \vdash k$ *can be packed into every partition of* n *with at most* r *parts. Then*

$$n(k_1, \ldots, k_t; r) = \max_{1 \leq i \leq qt} \left(\sum_{j=1}^{i} k_j + (k_i - 1)(r - 1) \right). \tag{2.4}$$

PROOF. We denote the righthand side of (2.4) by $f(k_1, \ldots, k_t; r)$. Clearly, $n(k_1, \ldots, k_t; r) \geq f(k_1, \ldots, k_t; r)$, since the packing $(k_1, \ldots, k_t) \subset (n - (r - 1)(k_i - 1), (k_i - 1)^{r-1})$ implies the inequality $n - (r - 1)(k_i - 1) \geq k_1 + \cdots + k_i$. Moreover, if $t > 1$, then

$$f(k_1, \ldots, k_t; r) \geq k_1 + f(k_2, \ldots, k_t; r), \tag{2.5}$$

because if i is an index maximizing $f(k_2, \ldots, k_t; r)$, then

$$f(k_1, \ldots, k_t; r) \geq \sum_{j=1}^{i} k_j + (r - 1)(k_i - 1) = k_1 + \sum_{j=2}^{i} k_j + (r - 1)(k_i - 1) =$$

$$= k_1 + f(k_2, \ldots, k_t; r).$$

We prove (2.4) by induction with respect to t. For $t = 1$ it is precisely Dirichlet's principle. For the induction step from $t - 1$ to t it suffices to prove that if $n = f(k_1, \ldots, k_t; r)$, then the required packability takes place. Consider an arbitrary partition $(n_1, \ldots, n_r) \vdash n$. We always have $n_1 \geq k_1$, since $f(k_1, \ldots, k_t; r) \geq k_1 + (r - 1)(k_1 - 1)$. Therefore the packing

$$(k_2, \ldots, k_t; r) \subset (n_1 - k_1, n_2, \ldots, n_r) \tag{2.6}$$

implies the packing $(k_1, \ldots, k_t) \subset (n_1, \ldots, n_r)$. In turn, (2.6) follows from (2.5) and the induction hypothesis

$$n_1 - k_1 + n_2 + \cdots + n_r = n - k_1 = f(k_1, \ldots, k_t; r) - k_1 \geq f(k_2, \ldots, k_t; r).$$

If, moreover, $n_1 = k_1$, then we can use the obvious monotonicity of f with respect to r. □

In the statement of the principle of complete packability the condition 'at most' can always be omitted, except in the degenerate case $k = t < r$. The principle can also be formulated in a dual manner, as a formula for the largest r for n fixed.

COROLLARY 2.2. *If* $k_1 > 1$, $k_1 \geq k_2 \geq \cdots \geq k_t$ *are natural numbers,* $(k_1, \ldots, k_t) \vdash k \leq n$, *and* $r(k_1, \ldots, k_t; n)$ *denotes the largest* r *such that the partition* $(k_1, \ldots, k_t)$ *can be packed into every partition of* n *into* r *parts, then*

$$r(k_1, \ldots, k_t; n) = \min_{i:\, k_i > 1} \left\lceil \frac{n - \sum_{j=1}^{i} k_j}{k_i - 1} \right\rceil + 1.$$

In fact, by the principle of complete packability the required r is the largest integer solution of the inequality $n \geq n(k_1, \ldots, k_t; n)$.

Facts concerning packability of partitions may be presented in terms of allocations. For example, if $k = n(k_1, \ldots, k_t; r)$, then each allocation of k particles over r boxes can be realized by allocating t groups of particles with k_j particles in group j ($j = 1, \ldots, t$), under the condition that each group is completely put into a single box. In particular, for each allocation of n particles over r boxes there are t distinct groups of particles (with $[(n + r - 1)/(t + r - 1)]$ particles in each group) that completely belong to the boxes. For example, Figure 2.1 makes clear that every allocation of 6 particles over two boxes can be realized by allocating only three compositions of 3, 2, and 1 particles, respectively. The question of the smallest number of such compositions which would realize all allocations of all particles is natural. Below we will consider this question in more detail.

It is obvious how to use the principle of complete packability to establish packability of two partitions: if we are to verify packability of a partition $(k_1, \ldots, k_t)$ into a partition $(n_1, \ldots, n_r)$, then fulfillment of the inequality $n \geq n(k_1, \ldots, k_t; r)$ implies the packing

$$(k_1, \ldots, k_t) \subseteq (n_1, \ldots, n_r).$$

For example, $n(3, 2, 1; 2) = \max(5, 5, 6) = 6$, and thus the partition $(3, 2, 1)$ can be packed into every partition of 6 with at most two parts. This remains true if we consider an arbitrary natural number not less than 6. Moreover, in this example Theorem 2.2 does not say anything about the packability of the partition $(4, 1, 1)$ into $(5, 1)$ or $(4, 2)$, nor on the packability of $(4, 1, 1)$ into $(3, 3)$.

5. Packability with restrictions

Sometimes it is required to guarantee a packability $(k_1, \ldots, k_t) \vdash k$ not in all partitions of a number n with r parts, but only in some of these. An extremal result guaranteeing packability of a fixed partition in not all partitions of a given rank is as follows.

THEOREM 2.3. *Let* $n_2, \ldots, n_r$, $k_1 \geq \cdots \geq k_t$, *and* r *be natural numbers, and let* $n(k_1, \ldots, k_t; n_2, \ldots, n_r)$ *be the smallest* n *for which every partition* $(p_1, \ldots, p_r)$ *of* n *with* r *parts and such that* $p_i \leq n_i$ ($i = 2, \ldots, r$) *has the property that* $(k_1, \ldots, k_t) \subseteq (p_1, \ldots, p_r)$. *Then*

$$n(k_1, \ldots, k_t; n_2, \ldots, n_r) = \max_{1 \leq i \leq t} \left(\sum_{j=1}^{i} k_j + \sum_{l=2}^{r} \min(n_l, k_i - 1) \right). \tag{2.7}$$

PROOF. a) If $f(k_1, \ldots, k_t)$ denotes the righthand side of (2.7), then

$$f(k_1, \ldots, k_t) - k_1 \geq f(k_2, \ldots, k_t; r). \tag{2.8}$$

In fact, if i maximizes $f(k_2, \ldots, k_t)$, then

$$f(k_1, \ldots, k_t) \geq \sum_{j=1}^{i} k_j + \sum_{l=2}^{r} \min(n_l, k_i - 1) =$$

$$= k_1 + \sum_{j=2}^{i} k_j + \sum_{l=2}^{r} \min(n_l, k_i - 1) = k_1 + f(k_2, \ldots, k_t).$$

b) Let $(p_1, \ldots, p_r)$ be a partition of $f(k_1, \ldots, k_t)$ with r parts in which $p_i \leq n_i$ $(i = 2, \ldots, r)$. Then it contains a part not less than k_1. In fact, in the opposite case we would have

$$f(k_1, \ldots, k_t) \geq k_1 + \sum_{l=2}^{r} \min(n_l, k_1 - 1) >$$

$$> \sum_{l=1}^{r} \min(p_l, k_1 - 1) = \sum_{l=1}^{r} p_l = f(k_1, \ldots, k_t).$$

c) We now prove the theorem by induction with respect to t. For $t = 1$ equation (2.7) gives

$$n(k; n_2, \ldots, n_r) = k + \sum_{l=2}^{r} \min(n_l, k - 1),$$

and by b), any partition $(p_1, \ldots, p_r)$ of $n(k; n_2, \ldots, n_r)$ with r parts and such that $p_i \leq n_i$ $(i = 2, \ldots, r)$ contains a part not less than k.

d) We now assume that the required holds up to $t - 1$ inclusive; we show it to hold for t too. Let $(p_1, \ldots, p_r)$ be an arbitrary partition of $f(k_1, \ldots, k_t)$ with r parts and such that $p_i \leq n_i$ $(i = 2, \ldots, r)$. Then by b), there is a part $p_j \geq k_1$. Thus, if

$$(k_2, \ldots, k_t) \subseteq (p_1, \ldots, p_j - k_1, \ldots, p_r), \tag{2.9}$$

then we have the packing $(k_1, \ldots, k_t) \subseteq (p_1, \ldots, p_r)$. In turn, (2.9) follows from (2.8) and the induction hypothesis

$$p_1 + \cdots + p_j - k_1 + \cdots + p_r = f(k_1, \ldots, k_t) - k_1 \geq$$

$$\geq f(k_2, \ldots, k_t) = n(k_2, \ldots, k_t; n_2, \ldots, n_r). \qquad \square$$

As distinct from the principle of complete packability, this theorem guarantees not only establishment of packability of partitions, but for certain partitions it can also be used as a nonpackability criterion. This is demonstrated in

COROLLARY 2.3. If $\sum_{i=1}^{r} n_i \geq n(k_1, \ldots, k_t; n_2, \ldots, n_r)$, then $(k_1, \ldots, k_t) \subseteq (n_1, \ldots, n_r)$.

If

$$n(k_1, \ldots, k_t; n_2, \ldots, n_r) > \sum_{i=1}^{r} n_i \geq \max_{2 \leq j \leq r} n(k_1, \ldots, k_t; n_2, \ldots, n_j - 1, \ldots, n_r),$$

then $(k_1, \ldots, k_t) \not\subseteq (n_1, \ldots, n_r)$.

PROOF. Let M be the set of all partitions $(p_1, \ldots, p_r)$ of the number $n = \sum_{i=1}^{r} n_i$ for which $p_i \leq n_i$ $(i = 2, \ldots, r)$, and let M_j be the subset of all partitions in M for which $p_j \leq n_j - 1$ $(j = 2, \ldots, r)$. Clearly, $M = \cup_{j=2}^{r} M_j \cup (n_1, \ldots, n_r)$. Since $n(k_1, \ldots, k_t; n_2, \ldots, n_r) > n$, there is in M a partition $(q_1, \ldots, q_r)$ such that $(k_1, \ldots, k_t) \nsubseteq (q_1, \ldots, q_r)$. However, since

$$n \geq \max_{2 \leq j \leq r} n(k_1, \ldots, k_t; n_2, \ldots, n_j - 1, \ldots, n_r),$$

for every $(p_1, \ldots, p_r) \in \cup_{j=2}^{r} M_j$ we have the packing $(k_1, \ldots, k_t) \subseteq (p_1, \ldots, p_r)$. Thus we must have $(q_1, \ldots, q_r) = (n_1, \ldots, n_r)$. $\square$

In particular, Theorem 2.3 thus allows us to assert that the partition $(4, 1, 1)$ can be packed into the partitions $(5, 1)$ and $(4, 2)$, while Corollary 2.3 establishes that it cannot be packed into the partition $(3, 3)$.

On the other hand, the relation between these theorems is very essential. More precisely, the function $n(k_1, \ldots, k_t; n_2, \ldots, n_r)$ is not only similar to the function $n(k_1, \ldots, k_t; r)$, but can sometimes be expressed in terms of it:

$$n(k_1, \ldots, k_t; (m^{r-1})) = \sum_{j=1}^{l} k_j + n(m + 1, k_{l+1}, \ldots, k_t; r) - m - 1,$$

where $k_{l+1} \leq m < k_l$. In fact,

$$n(k_1, \ldots, k_t; (m^{r-1})) = \max_{1 \leq i \leq t} \left(\sum_{j=1}^{i} k_j + (r - 1) \min(m, k_i - 1) \right) =$$

$$= \max \left(\sum_{j=1}^{l} k_j + (r - 1)m, \sum_{j=1}^{l} k_j + n(k_{l=1}, \ldots, k_t; r) \right) =$$

$$= \sum_{j=1}^{l} k_j + \max((r - 1)m, n(k_{l+1}, \ldots, k_t; r)) =$$

$$= \sum_{j=1}^{l} k_j + n(m + 1, k_{l+1}, \ldots, k_t; r) - m - 1.$$

6. Extrema of complete packability

We call the quantity $n(k_1, \ldots, k_t; r)$ the *bound of complete packability* and the righthand side of (2.1) the *function of complete packability*. We denote the latter by $f(x_1, \ldots, x_t; r)$, or by $f(X)$ if the value of r is clear from the context. Corollary 2.1 is equivalent to the equality

$$\max_{(k_1, \ldots, k_t) \vdash k} n(k_1, \ldots, k_t; r) = \max(k, r(k - t) + 1).$$

This value is realized for $(k_1, \ldots, k_t) = (k - t + 1, 1^{t-1})$.

One of the main questions investigated here consists of estimating the smallest possible value of a bound of complete packability, i.e. to compute the quantity

$$m(k, t, r) = \min_{(k_1, \ldots, k_t) \vdash k} n(k_1, \ldots, k_t; r).$$

Before passing to this estimation, we note some simplest properties of the function and bound of complete packability.

The monotonicity of $f(X;r)$ with respect to r is characterized in the following way. Let $r_1 \geq r_2$, and let v and w be indices maximizing $f(X;r_1)$ and $f(X;r_2)$, respectively. Then

$$(r_1 - r_2)(x_v - 1) \geq f(X;r_1) - f(X;r_2) \geq (r_1 - r_2)(x_w - 1).$$

In fact,

$$(r_1 - r_2)(x_v - 1) = \sum_{j=1}^{v} x_j + (r_1 - 1)(x_v - 1) - \sum_{j=1}^{v} x_j - (r_2 - 1)(x_v - 1) \geq$$

$$\geq f(X;r_1) - f(X;r_2) \geq$$

$$\geq \sum_{j=1}^{w} x_j + (r_1 - 1)(x_w - 1) - \sum_{j=1}^{w} x_j - (r_2 - 1)(x_w - 1) \geq (r_1 - r_2)(x_w - 1).$$

In particular, this monotonicity characterizes the subset of indices from $[t] = \{1,\ldots,t\}$ that maximize $f(X;r)$.

Let $I(r) = \{i \in [t]:\ f(X;r) = \sum_{j=1}^{i} x_j + (r - 1)(x_i - 1)\} \subseteq [t]$. Then $|I(r) \cap I(r + 1)| \leq 1$. Moreover, if $a(r) = \min_{i \in I(r)} i$, $b(r) = \max_{i \in I(r)} i$, then $b(r) \leq a(r-1)$, which, in turn, gives a recurrent way of computing the function of complete packability, as follows:

$$f(X;r) = \max_{1 \leq i \leq a(r-1)} \left(\sum_{j=1}^{i} x_j + (r - 1)(x_i - 1) \right),$$

where $a(1) = 1$.

If X and Y are real vectors of identical dimension, then

$$f(X + Y;r) \leq f(X;r) + f(Y;r) + r - 1. \tag{2.10}$$

In fact, suppose the indices u, v, and w maximize the functions $f(X+Y;r)$, $f(X;r)$, and $f(Y;r)$, respectively. Then

$$f(X;r) + f(Y;r) + r - 1 = \sum_{j=1}^{v} x_j + (r-1)(x_v - 1) + \sum_{j=1}^{w} y_j + (r-1)(y_w - 1) + r - 1 \geq$$

$$\geq \sum_{j=1}^{u}(x_j + y_j) + (r - 1)(x_u + y_u - 1) = f(X + Y;r).$$

For l a natural number, the following equations hold:

$$f(lk_1,\ldots,lk_t;r) = lf(k_1,\ldots,k_t;r) + (r - 1)(l - 1), \tag{2.11}$$

$$f(k_1^l,\ldots,k_t^l;rl - l + 1) = lf(k_1,\ldots,k_t;r), \tag{2.12}$$

$$f(k_1^l,\ldots,k_t^l;r) = lf\left(k_1,\ldots,k_t;1 + \frac{r-1}{l}\right). \tag{2.13}$$

The same equations hold for the bound of complete packability, for the same parameters for which the quantity $n(k_1,\ldots,k_t;r)$ is well defined.

What is the meaning of these formulas when l is not a natural number? One experiences a strange sensation when looking at the righthand side of (2.13): we are already able to visualize multisets in which the elements have nonnatural multiplicities, but how could we comprehend a partition with a nonnatural number of parts? However, these formulas work quite efficiently (which may be due to its level of formalization) and make it possible to compute certain conditional extrema. For example, if we are to compute the minimum of the function of complete packability,

$$\mu(k,t,r) = \min_{(k_1,\ldots,k_t) \vdash k} f(k_1,\ldots,k_t;r),$$

not over all vectors $(k_1,\ldots,k_t) \vdash k$ but only over those in which each component is present exactly l times, then, denoting this minimum by $\mu_l(lk,lt,r)$, i.e. putting

$$\mu_l(lk,lt,r) = \min_{(k_1^l,\ldots,k_t^l)\vdash k} f(k_1^l,\ldots,k_t^l;r),$$

using the inequality (2.13) we find

$$\mu_l(lk,lt,r) = l\mu\left(k,t,1+\tfrac{r-1}{l}\right).$$

For $X \in \mathbf{N}^t$ the function $n(X;r)$ can be defined for $r \in \mathbf{R}$ by $n(X;r) = f(X;r)$. This definition is correct, since $f(X;r)$ is continuous and monotone nondecreasing with respect to r.

The exact values of certain extrema of the function of complete packability are given by

LEMMA 2.4. *Let* $X = (x_1,\ldots,x_t) \in \mathbf{R}^t$. *Then*

1) *if* $\sum_{i=1}^t x_i = p \geq 0$ *and* $f(X;r) = \sum_{j=1}^i x_j + (x_i-1)(r-1)$, *then* $x_i \geq 0$;
2) *if* $x_1 \geq \cdots \geq x_l > 0 \geq x_i$, $i = l+1,\ldots,t \geq l \geq 1$, *and* S_t *is the symmetric group of all permutations of the set* $[t] = \{1,\ldots,t\}$, *then*

$$\max_{s \in S_t} f(x_{s(1)},\ldots,x_{s(t)};r) = \sum_{j=1}^l x_j + (x_1-1)(r-1); \qquad (2.14)$$

if, moreover, $\sum_{i=1}^t x_i \geq 0$, *then*

$$\min_{s \in S_t} f(x_{s(1)},\ldots,x_{s(t)};r) = \sum_{j=l+1}^t x_j + f(x_1,\ldots,x_l;r); \qquad (2.15)$$

3) *if* $\sum_{i=1}^t x_i = 0$, $\sum_{i=1}^t |x_i| = d$, *then*

$$\max_X f(X;r) = 1 - r + \tfrac{dr}{2}, \qquad (2.16)$$

$$\min_X f(X;r) = \frac{d(r-1)^{t-1}}{2(r^{t-1} - (r-1)^{t-1})} - r + 1; \qquad (2.17)$$

4) *if* $\sum_{i=1}^t x_i = p \in \mathbf{R}$ *and* $1 \leq r \in \mathbf{R}$, *then*

$$\min_X f(X;r) = \frac{pr^t}{r^t - (r-1)^{t-1}} - r + 1. \qquad (2.18)$$

PROOF. 1) Assume the contrary: $x_i < 0$, and consider the largest $m \in [i-1]$ for which $x_m \geq 0$. Then

$$\sum_{j=1}^{m} x_j + (x_m - 1)(r-1) > \sum_{j=1}^{i} x_j + (x_i - 1)(r-1), \qquad (2.19)$$

since $x_{m+1}, \ldots, x_i$ are negative. If such an m does not exist, then we consider the largest $m \in [i+1, t] = [t] \setminus [i]$ for which $x_m \geq 0$. But then (2.19) is again true, since $p \geq 0$ and $x_1, \ldots, x_i, x_m, \ldots, x_t$ are all negative. In fact, in this case (2.19) is equivalent to the inequality

$$\sum_{j=i+1}^{m} x_j + (x_m - 1)(r-1) > (x_i - 1)(r-1),$$

which follows from the fact that $x_m \geq 0 > x_i$ and $\sum_{j=i+1}^{m} x_i \geq 0$, since $p \geq 0$ and the last sum is over all positive x_j.

2) $\sum_{j=1}^{l} x_j + (x_1 - 1)(r-1) = f(x_l, \ldots, x_1, x_{l+1}, \ldots, x_t) \leq \max_{s \in S_t} f(X_s) \leq \max_s \max_i \sum_{j=1}^{i} x_{s(i)} + \max_s \max_i (x_{s(i)} - 1)(r-1) = \sum_{j=1}^{l} x_j + (x_1 - 1)(r-1)$. This proves (2.14).

To prove (2.15) we note the following two facts.

a) We first show that there is a permutation of the form $(x_{l+1}, \ldots, x_t, x_{s(1)}, \ldots, x_{s(l)})$ minimizing f. For this it suffices to verify that if for an arbitrary permutation $Y = (y_1, \ldots, y_t)$, $y_q < 0$, then for $Y' = (y_q, y_1, \ldots, y_{q-1}, y_{q+1}, \ldots, y_t)$ we have $f(Y') \leq f(Y)$.

In fact, if

$$f(Y') = y_q + \sum_{\substack{j=1 \\ j \neq q}}^{p} y_j + (y_p - 1)(r-1),$$

then

$$f(Y) = \sum_{j=1}^{p} y_j + (y_p - 1)(r-1) \geq y_q + \sum_{\substack{j=1 \\ j \neq q}}^{p} y_j + (y_p - 1)(r-1) = f(Y'),$$

and since $\sum_{j=1}^{t} y_j \geq 0$, by 1) above we have $p \neq q$.

b) We now prove that if $x_1 \geq \cdots \geq x_l$, then

$$\min_{s \in S_l} f(x_{s(1)}, \ldots, x_{s(l)}; r) = f(x_1, \ldots, x_l; r).$$

For this we verify that if $x_i \leq x_{i+1}$, then

$$f(x_1, \ldots, x_l; r) \geq f(x_1, \ldots, x_{i+1}, x_i, \ldots, x_l; r).$$

In fact, we put

$$X = (x_1, \ldots, x_l), \qquad X' = (x_1, \ldots, x_{i+1}, x_i, \ldots, x_l).$$

Then: if $f(X') = \sum_{j=1}^{p} x_j + (x_p - 1)(r-1)$, where $p \neq i, i+1$, then

$$f(X) \geq \sum_{j=1}^{p} x_j + (x_p - 1)(r-1) = f(X');$$

if $f(X') = \sum_{j=1}^{i-1} x_j + x_i + (x_{i+1} - 1)(r - 1)$, then

$$f(X) \geq \sum_{j=1}^{i+1} x_j + (x_{i+1} - 1)(r - 1) \geq \sum_{j=1}^{i+1} x_j + x_{i+1} + (x_{i+1} - 1)(r - 1) = f(X');$$

if $f(X') = \sum_{j=1}^{i+1} x_j + (x_i - 1)(r - 1)$, then

$$f(X) \geq \sum_{j=1}^{i+1} x_j + (x_{i+1} - 1)(r - 1) \geq \sum_{j=1}^{i+1} x_j + +(x_i - 1)(r - 1) = f(X').$$

Since every permutation can be represented as a product of transpositions, b), and thus (2.15), has been proved. In particular, (2.15) implies that if X is an integral vector with positive components, then

$$\min_{s \in S_t} f(X_s; r) \geq \sum_j x_j. \tag{2.20}$$

3) $1 - r + dr/2 = f(d/2, 0, \ldots, 0, -d/2; r) \leq \max f(X; r) \leq \max_i \sum_{j=1}^i x_i + \max_i (x_i - 1)(r - 1) \leq d/2 + (d/2 - 1)(r - 1) = 1 - r + dr/2$. Thus (2.16) has been proved.

Equality in (2.17) is realized for

$$Z = \left(-\frac{d}{2}, \frac{dr^{t-2}}{2(r^{t-1} - (r-1)^{t-1})}, \frac{dr^{t-3}(r-1)}{2(r^{t-1} - (r-1)^{t-1})}, \ldots, \frac{d(r-1)^{t-2}}{2(r^{t-1} - (r-1)^{t-1})} \right).$$

We show that (2.17) holds as a lower bound. Let $f(X) = \sum_{j=1}^i x_j + (x_i - 1)(r - 1)$. By 1) above, $x_i \geq 0$, and then

$$f(X) \geq -\frac{d}{2} + \min_{\substack{x_j: \sum x_j = \frac{d}{2} \\ x_j \geq 0}} f(x_1, \ldots, x_l; r) = -\frac{d}{2} + \frac{dr^l}{2(r^l - (r-1)^l)} - r + 1 =$$

$$= \frac{d(r-1)^l}{2(r^l - (r-1)^l)} - r + 1 \geq \frac{d(r-1)^{t-1}}{2(r^{t-1} - (r-1)^{t-1})} - r + 1.$$

Here, the first inequality holds because of (2.15) and the equality following it because of (2.18), which proves (2.17).

4) Let $q_i = \frac{pr^{t-i}(r-1)^{i-1}}{r^t - (r-1)^t}$ $(i = 1, \ldots, t)$; $Q = (q_1, \ldots, q_t)$. Then $\sum_{i=1}^t q_i = p$ and

$$f(Q; r) = \frac{pr^t}{r^t - (r-1)^t} - r + 1, \tag{2.21}$$

and for each $i \in [t]$ the equation $\sum_{j=1}^i q_j + (q_i - 1)(r - 1) = f(Q; r)$ holds. For $X = (x_1, \ldots, x_t)$, $\sum_{i=1}^t x_i = p$, the latter implies

$$f(X; r) = \frac{pr^t}{r^t - (r-1)^t} + f(X - Q; r). \tag{2.22}$$

Moreover,

$$\min_{X: \sum x_i = 0} f(X; r) = f(0; r) = -r + 1. \tag{2.23}$$

In fact, assuming the contrary, i.e. $\min f < -r + 1$, we are led to the system

$$\sum_{j=1}^{t} x_j = 0,$$

$$\sum_{j=1}^{i} x_j + r x_i < 0, \qquad i = 1, \ldots, t,$$

which is contradictory for $r \geq 1$. Indeed, subtracting the equality from the t inequalities we obtain $(r-1)x_i < x_{i+1} + \cdots + x_t$ $(i = 1, \ldots, t)$, whence we successively find that $x_t < 0$, $x_{t-1} < 0, \ldots, x_1 < 0$, contradicting the equality in this system. To prove (2.23) we cannot refer to (2.17) with $d = 0$, since in the proof of (2.17) we have used (2.18). Using now (2.21), (2.22) for $k = p$ and (2.23), we obtain the required. $\square$

COROLLARY 2.4. *If $k \equiv 0$ (mod $r^t - (r-1)^t$), then*

$$m(k, t, r) = \frac{kr^t}{r^t - (r-1)^t} - r + 1; \tag{2.24}$$

this value is attained for the partition

$$\left(\frac{kr^{t-i}(r-1)^{i-1}}{r^t - (r-1)^t} \right), \qquad 1 \leq i \leq t.$$

We now consider the estimation of $m(k, t, r)$ when $k \not\equiv 0$ (mod $r^t - (r-1)^t$). Everywhere below, Q denotes the t-dimensional vector with components

$$q_i = \left(\frac{kr^{t-i}(r-1)^{i-1}}{r^t - (r-1)^t} \right), \qquad i = 1, \ldots, t,$$

and $l = \sum_{i=1}^{t}\{q_i\}$. We note the following property of Q: all components q_i are either all nonintegers or all integers; the latter case holds if and only if k is a multiple of $r^t - (r-1)^t$. This follows directly from the fact that the numbers $r^{t-i}(r-1)^{i-1}$ and $r^t - (r-1)^t$ are coprime for every $i \in [t]$. Consequently, for every $i \in [t]$ we have

$$\frac{1}{r^t - (r-1)^t} \leq \{q_i\} \leq 1 - \frac{1}{r^t - (r-1)^t}. \tag{2.25}$$

Moreover, it is obvious that $1 \leq l \leq t - 1$.

We now prove the following lower bound:

$$m(k, t, r) \geq \frac{kr^t}{r^t - (r-1)^t} - r + 1 + l - f(\{Q\}; r), \tag{2.26}$$

where $\{Q\} = (\{q_1\}, \ldots, \{q_t\})$.

If $K = (k_1, \ldots, k_t) \vdash k$, then by (2.22) we have

$$f(K; r) = \frac{kr^t}{r^t - (r-1)^t} + f(K - Q; r).$$

In turn,

$$f(K - Q; r) = f(K - [Q] - \{Q\}; r) \geq f(K - [Q]; r) - f(\{Q\}; r) - r + 1.$$

Here, $[Q] = ([q_1], \ldots, [q_t])$ and $[q]$ is the integer part of the number $[q]$. The latter inequality follows from (2.10) by putting $X = K - [Q] - \{Q\}$, $Y = \{Q\}$. Moreover, by the remark accompanying (2.20) we have $f(K - [Q]; r) \geq \sum_{i=1}^{t}(k_i - [q_i]) = l$. Thus,

$$f(K - [Q]; r) - f(\{Q\}; r) - r + 1 \geq l - r + 1 - f(\{Q\}; r).$$

Note that the righthand side of (2.26) is always an integer. The bound given implies that if $\{q_M\} = \max_{1 \leq i \leq t}\{q_i\}$, then

$$m(k, t, r) \geq \frac{kr^t}{r^t - (r-1)^t} - \{q_M\}(r-1),$$

since

$$f(\{Q\}; r) \leq \max_{1 \leq i \leq t} \sum_{j=1}^{i}\{q_j\} + \max_{1 \leq i \leq t}(\{q_i\} - 1)(r-1) \leq l + (\{q_M\} - 1)(r-1).$$

Another lower bound is of more geometrical shape; more precisely, if $d = \min\|Z - Q\|_{\ell_1}$, where the min is taken over all integer vectors $Z = (z_1, \ldots, z_t)$ such that $\sum_{i=1}^{t} z_i = k$ and ℓ_1 is the space of infinite number sequences with finite sum of absolute values, then

$$m(k, t, r) \geq \frac{kr^t}{r^t - (r-1)^t} + \frac{d(r-1)^{t-1}}{2(r^{t-1} - (r-1)^{t-1})} - r + 1.$$

This bound follows immediately from (2.17) and (2.22). In particular, it suggests that the minimizing partition cannot be an integral vector that lies too far away from Q in the metric of ℓ_1. Moreover, it may be used for estimating the bound of complete packability of any concrete partition $K = (k_1, \ldots, k_t)$;

$$n(K; r) \geq \frac{kr^t}{r^t - (r-1)^t} + \frac{\|K - Q\|_{\ell_1}(r-1)^{t-1}}{2(r^{t-1} - (r-1)^{t-1})} - r + 1.$$

As upper bound we prove the inequality

$$\frac{kr^t}{r^t - (r-1)^t} - \{q_{ml}\}(r-1) \geq m(k, t, r),$$

where $\{q_{ml}\} = \min_{t-l+1 \leq i \leq t}\{q_i\}$. Consider the partition

$$K = ([q_1], \ldots, [q_{t-l}],]q_{t-l+1}[, \ldots,]q_t[) \vdash k.$$

By 1) above, for it we have

$$f(K - Q) = \max_{t-l+1 \leq i \leq t}\left(\sum_{j=1}^{i}(k_j - q_j) + (k_i - q_i - 1)(r-1)\right) \leq$$

$$\leq \max_{t-l+1 \leq i \leq t}\sum_{j=1}^{i}(k_j - q_j) + \max_{t-l+1 \leq i \leq t}(k_i - q_i - 1)(r-1) = -\min_{t-l+1 \leq i \leq t}\{q_i\}(r-1).$$

Taking into account (2.22) this implies the required.

Thus, (2.25), (2.26) and the last estimate imply the two-sided bound

$$\frac{kr^t - r + 1}{r^t - (r-1)^t} \geq m(k, t, r) \geq \frac{kr^t + r - 1}{r^t - (r-1)^t} - r + 1. \tag{2.27}$$

We consider some refinements of the general bounds obtained.

For $r \leq 2$ the difference between the upper and lower bounds is always less than one, and, since $m(k, t, r)$ is an integer, they provide an exact value:

$$m(k, t, r) = \left\rceil \frac{kr^t + r - 1}{r^t - (r-1)^t} \right\rceil - r + 1, \qquad r = 1, 2. \tag{2.28}$$

PROPOSITION 2.1. *If* $k, t, p \in \mathbf{N}$ *and* $1 \leq r \leq 2$, *then*

$$m(k, t, r) = \left\rceil \frac{pk(p + r - 1)^t}{(p + r - 1)^t - (r-1)^t} \right\rceil + 1;$$

moreover, if

$$K = \left(\left[\frac{pk(p + r - 1)^{t-1}(r - 1)^0}{(p + r - 1)^t - (r-1)^t} \right], \dots, \left[\frac{pk(p + r - 1)^l(r - 1)^{t-l+1}}{(p + r - 1)^t - (r-1)^t} \right], \right.$$
$$\left. \left\rceil \frac{pk(p + r - 1)^{l-1}(r - 1)^{t-l}}{(p + r - 1)^t - (r-1)^t} \right[, \dots, \left\rceil \frac{pk(p + r - 1)^0(r - 1)^{t-1}}{(p + r - 1)^t - (r-1)^t} \right[\right),$$

where

$$l = \sum_{i=1}^{t} \left\{ \frac{pk(p + r - 1)^{t-i}(r - 1)^{i-l}}{(p + r - 1)^t - (r-1)^t} \right\},$$

then $n(K^p; r) = m_p(pk, pt; r)$.

(For a vector K the notation K^p denotes the vector with each component repeated p times.)

PROOF.

$$m_p(pk, pt; r) \geq \mu_p(pk, pt; r) = p\mu(k, t; 1 + (r-1)/p) =$$
$$= p \left(\frac{k(p + r - 1)^{tl}}{(p + r - 1)^t - (r-1)^t} - \frac{r-1}{p} \right) = \frac{pk(p + r - 1)^t}{(p + r - 1)^t - (r-1)^t} - r + 1,$$

which implies the required lower bound since m_p is an integer. On the other hand,

$$n(K^p; r) = pn \left(K; 1 + \frac{r-1}{p} \right) < p \frac{k(p + r - 1)^t}{(p + r - 1)^t - (r-1)^t};$$

consequently, the difference between the upper and lower bounds is strictly less than one if $r \leq 2$.

For $l = 1$ we use the partition

$$K = ([q_1], \dots, [q_{M-1}],]q_M[, [q_{M+1}], \dots, [q_t]) \vdash k$$

in the upper bound, where $\{q_M\} = \max_{1 \leq i \leq t}\{q_i\}$. By Lemma 2.4, 1), we have

$$f(K - Q) = - \sum_{j=1}^{M-1} \{q_j\} + (1 - \{q_M\}) - \{q_M\}(r - 1) = 1 - \sum_{j=1}^{M} \{q_j\} - \{q_M\}(r - 1) =$$

$$= \sum_{j=M+1}^{t} \{q_j\} - \{q_M\}(r - 1) < 1 - \{q_M\}(r - 1).$$

In comparison with the lower bound in (2.26) this gives a difference less than one. Consequently, if $l = 1$,

$$m(k, t; r) = \frac{kr^t}{r^t - (r-1)^t} + \sum_{j=M+1}^{t} \{q_j\} - \{q_M\}(r-1), \qquad (2.29)$$

or

$$m(k, t; r) = \left] \frac{kr^t}{r^t - (r-1)^t} - \{q_M\}(r-1) \right[. \qquad (2.30)$$

Thus, for $l = 1$ the bound of complete packability minimizes the integral vector which is closest to Q in the metric of ℓ_1. In particular, this gives a complete solution of the minimization problem for $t = 2$. In the general case this phenomenon does not always hold; the smallest (with respect to t) example known is $k = 422$, $t = 5$, $r = 5$. For these parameters $Q = (125.5355, 100.4284, 80.3427, 64.27416, 51.41933)$. The integral vector nearest to Q in the metric of ℓ_1 is $X = (126, 101, 80, 64, 51)$; for it $\|X - Q\|_{\ell_1} = 2.072346$ and $n(X; 5) = 627$. However, for the vector $Y = (126, 100, 80, 64, 52)$ we have $\|Y - Q\|_{\ell_1} = 2.090435$ and $n(Y; 5) = 626$. The smallest (with respect to k and r) example known is given by the parameters $k = 76$, $t = 6$, $r = 3$. For it, $Q = (27.77143, 18.51429, 12.34286, 8.228571, 5.485715, 3.657143)$. The integral vector nearest to Q in the metric of ℓ_1 is $X = (28, 19, 12, 8, 5, 4)$, for which $\|X - Q\|_{\ell_1} = 2.114285$ and $n(X; 3) = 83$. However, for $Y = (28, 18, 12, 8, 6, 4)$ we have $\|Y - Q\|_{\ell_1} = 2.171428$ and $n(Y; 3) = 82$. $\square$

THEOREM 2.4. *The quantity $m(k, t, r)$ is the smallest integer C such that the recursively defined vector*

$$y_i = \left[\frac{C + r - 1 - \sum_{j=1}^{i-1} y_j}{r} \right], \qquad i = 1, \ldots, t,$$

satisfies the equation

$$\sum_{j=1}^{t} y_j = k.$$

We immediately note that by the above given bounds for $m(k, t, r)$, the possible choice of C does not exceed the difference between these bounds, i.e. $r - 1$. Since the sum $\sum_{j=1}^{t} y_j$ is monotone with respect to C, this choice can be realized by operations whose number is $\log_2 r$.

PROOF. Clearly, always $n(Y; r) \leq C$, therefore we can put $C = m(k, t, r)$. If, moreover, $\sum_{j=1}^{t} y_j = k$, then $n(Y; r) = m(k, t, r)$ (since $m(k, t, r)$ is minimal). If $\sum_{j=1}^{t} y_j > k$, then, by reducing the necessary amount of components y_j of Y by $\sum_{j=1}^{t} y_j - k$, we obtain a partition of k having bound of complete packability at most, and hence equal to, $m(k, t, r)$.

In the alternative case $\sum_{j=1}^{t} y_j < k$ we consider a minimizing partition $X = (x_1, \ldots, x_t) \vdash k$ which is 'largest' in the sense of the lexicographical ordering $\succeq$ on $P_t(k)$, and also consider the smallest $i \in [t]$ for which $x_i > y_i$. Clearly, such an i exists always, and $[i - 1] \neq \emptyset$ since $x_1 \leq [(m + r - 1)/r] = y - 1$.

There is a $j \in [i-1]$ such that $x_j < y_j$, since otherwise ($x_j = y_j$, $j = 1, \ldots, i-1$, $x_i > y_i$) we would have $\sum_{j=1}^{i} x_j + (r-1)(x_i - 1) > m(k, t, r)$. In fact, since in this case $x_i \geq y_i + 1$, we would have

$$\sum_{j=1}^{i} x_j + (r-1)(x_i - 1) \geq \sum_{j=1}^{i} y_j + (r-1)(y_i - 1) + r > m(k, t, r),$$

because nonsatisfaction of the last inequality is equivalent to

$$y_i \leq \left[\frac{m - 1 - \sum_{j=1}^{i-1} y_j}{r}\right] = y_i - 1.$$

We now consider a $j \in [i-1]$ such that $x_j < y_j$, and estimate the bound of complete packability, $n(X'; r)$, for the partition

$$X' = (x_1', \ldots, x_t') = (x_1, \ldots, x_j + 1, \ldots, x_i - 1, \ldots, x_t) \vdash k.$$

Let v be an index maximizing $n(X'; r)$, i.e.

$$n(X'; r) = \sum_{w=1}^{v} x_w' + (r-1)(x_v' - 1).$$

If $v > i$, then clearly $n(X'; r) = n(X; r) = m(k, t, r)$.
   If $v = i$, then

$$n(X'; r) = \sum_{w=1}^{i} x_w' + (r-1)(x_i' - 1) = \sum_{w=1}^{i} x_w + (r-1)(x_i - 2) + 1 =$$

$$= \sum_{w=1}^{i} x_w + (r-1)(x_i - 1) - r + 2 \leq n(X; r) - r + 2 = m(k, t, r) - r + 2,$$

which does not exceed $m(k, t, r)$ for $r \geq 2$.
   If $v < i$, then $x_w' \leq y_w$ ($w = 1, \ldots, v$). Thus, in this case too,

$$n(X'; r) = \sum_{w=1}^{v} x_w' + (r-1)(x_v' - 1) \leq n(Y; r) \leq m(k, t, r).$$

Thus, $n(X'; r) \leq m(k, t, r)$, and hence $n(X'; r) = m(k, t, r)$, i.e. the partition X' is also minimizing. At the same time $X' \succeq X$, contradicting the maximality of X in the sense of the lexicographical ordering. $\square$

Let $t(k, r)$ be the smallest t for which there is a partition of k with t parts that can be packed into every partition of k with r parts.

COROLLARY 2.5. $t(k, r)$ is the smallest t such that the recursively defined vector

$$y_i = \left[\frac{k + r - 1 - \sum_{j=1}^{i-1} y_i}{r}\right], \qquad i = 1, \ldots, t,$$

satisfies the equation

$$\sum_{j=1}^{t} y_j = k.$$

In fact, this t is the smallest root of the equation $k = m(k, t, r)$. Here, the choice is again restricted by bounds available for t. These bounds follow from those for $m(k, t, r)$:

$$\frac{\ln(k + r - 1)}{\ln(r) - \ln(r - 1)} \geq t(k, r) \geq \frac{\ln(k + r - 1) - \ln(r - 1)}{\ln(r) - \ln(r - 1)}.$$

In fact, Theorem 2.4 does not only give the exact value of $m(k, t, r)$, but also bounds $t(k, r)$. More precisely, if $t \geq t(k, r)$, then the number of nonzero components of the vector Y in the statement of Theorem 2.4 is exactly $t(k, r)$. Clearly, if $t \geq t(k, r)$, then $m(k, t, r) = k$. Thus, for $t > t(k, r)$ this vector Y must have nonzero components. If in this case it is necessary to construct the partition $p \in P_t(k)$ minimizing the bound of complete packability $n(p; r) = k$, then we can clearly take p to be any partition in $P_t(k)$ which can be packed into the vector Y, regarded as a partition. The lexicographically largest such partition can also be given recursively, and has, for $i = 1, \dots, t$, the form

$$k_i = \min\left(\left[\frac{k + r - 1 - \sum_{j=1}^{i-1} k_j}{r}\right], k - \sum_{j=1}^{i-1} k_j - t + i\right).$$

7. Weighing

'Every sensitive body, of whatever shape, placed on balances in equilibrium with weights, in all positions absolutely retains some equilibrium .... One often neglects or even rushes by simple and everyday phenomena which in the examination of nature give rein to great revelations, and one pre-accepts and searches for difficult tests, forgetting the once famous example, i.e. the simplicity and undoubtful mathematical fundament that every thing equals itself by quantity, to which almost all of mathematics appeals'.

Such wrote M.V. Lomonosov in its work 'Discussions on the rigidity and wateriness of bodies'. Here we turn to this 'famous example' of human activity: weighing.

If, on a balance weight, we place a load of weight n in one pan and balance it by weights $n_1, \dots, n_r$ in the other pan, then we have $n = n_1 + \cdots + n_r$. Since balance 'in all positions absolutely retains some equilibrium', the order of the terms in this sum is unimportant in essence. Thus, the latter sum must be some partition of n.

We now present ourselves with the real situation that we have to weight a load V by weights $k_1, \dots, k_t$ with total weight $k_1 + \cdots + k_t \geq V$, and we describe this situation analytically. If V is balanced by some of the k_i, then

$$V = k_{i_1} + \cdots + k_{i_l},$$

where $1 \leq i_1 < \cdots < i_l \leq t$. Of course, we do not necessarily need all k_i ($1 \leq i \leq t$) to balance V. But what about the remaining weights $k_{i_{l+1}}, \dots, k_{i_t}$, i.e. those not taking part in balancing V? It is obvious that they also balance some load, to be precise $k - V$, i.e. together with the identity above another identity holds:

$$k - V = k_{i_{l+1}} + \cdots + k_{i_t}.$$

So, the system of equations thus obtained is a *packing* of the partition $(k_1, \ldots, k_t)$ into the partition $(V, k - V)$. Consequently, this apt relation between weighing and partitions and packability leads to the following facts.

FACT 2.1. A system of weights $k_1, \ldots, k_t$ of total weight k balances a load V exactly if and only if there is a packing $(k_1, \ldots, k_t) \subseteq (V, k - V)$.

FACT 2.2. A system of weights $k_1, \ldots, k_t$ of total weight k balances any integer load not heavier than k if and only if the partition $(k_1, \ldots, k_t)$ can be packed into every partition of k having two parts.

We stress the meaning of these facts: in them real engineering phenomena are adequately described by only the abstract mathematical notions of number partitions and the binary relation of packability of partitions.

The notions of partition and packability not only adequately describe an equilibrium state, but also turn out to be useful in solving many weighing problems, among which we take as initial one the following old problem.

Problem. How many weights does one minimally need to weigh any integer number of pounds from 1 to k?

For $k = 40$ this problem is stated in 'The Book of Abacus' (1202 A.D.) by the Italian merchant and mathematician Leonardo Pisano, whose nickname was Fibonacci (which means 'Son of Bonacco').

Fibonacci travelled for a long time in the Orient, where he met with the Arab mathematicians Al-Khwarizmi, Abu Kamil, Mohammed-ben-Muza. In turn, they drew his attention to the mathematicians of India, notably Brahmagupta. The works 'The Book of Abacus', 'The Practice of Geometry', 'Algebra', 'Treatise on Square Numbers' kept up and nourished the European mathematicians of the Middle Ages. They played an essential role in spreading the Arab ciphers. Thus, Leonardo reported on how one could write any number by using the 9 ciphers and zero, called the Arab ciphers.

The name of zero, al zyfr (tefiro, tefro, zero), gave also the name of the signs themselves: zifrae, eifrae.

Under his nickname he became immortal for the Fibonacci sequence $1, 1, 2, 3, 5, 8, 13, \ldots$, in which the sum of two subsequent terms gives the next term. This sequence, which characterizes in Leonardo's original treatise the fertility of rabbits, turned out to be far more fertile in mathematics than the phenomenon first described by it.

Another Italian mathematician was drawn to this problem, Lucas Paccioli (1445–1514). The inscription on his tombstone in Sansepolgro, erected in 1878, reads:

'To Lucas Paccioli, who was a friend and contemporary of Leonardo da Vinci and Leon Batista Alberti, who first gave algebra its language and the structure of a science, who applied his great discoveries to geometry, invented double bookkeeping, and gave in mathematical treatises the basics and invariable norms for subsequent investigations'

This list can be completed by yet another achievement: Paccioli undertook the first attempts at a geometrical modeling of writing. The well-known harmonic square of Leonardo da Vinci contains a human being; the harmonic square of Paccioli contains a letter.

Paccioli proposed (1509) to construct letters on the basis of a square (each side of which was partitioned into nine parts).

This problem is also encountered in the book 'Applied and Remarkable Problems on Numbers', Lyon, 1612. Its author, the Frenchman Claude Gaspar Bachet de Méziriac (1587–1638), was a remarkable person in many respects: poetry, mathematics, translator, publisher and commentator of the 'Arithmetic' of Diophantus of Alexandrië, the same edition in which the 'margin was too small' to contain the proof of the great Fermat theorem. The Russian translation of Bachet's book 'Games and Problems Based on Mathematics' appeared in St. Petersburg in 1877. Bachet had a preference for problems of Antiquity, as well as for those based on some real situations. The weighing problem is sometimes termed *Bachet's problem*. A propos, it is Bachet to whom we owe a present-day text of the description of life in Aesop.

An analytic approach to similar problems was proposed by Leonard Euler (1707–1783). His approach is based on the method of generating functions developed by him. In his 'Analysis Infinitorum', more precisely in the chapter on number partitions, Euler concludes from the easily established analytic identity

$$(1 + x)(1 + x^2)(1 + x^4)(1 + x^8) \cdots = 1 + x + x^2 + x^3 + x^4 + x^5 + \ldots$$

'that every [nonnegative integral] number can be obtained from various terms of the double [i.e. with multiplier 2] geometrical progression $1, 2, 4, 8, 16, 32, \ldots$ by addition, and moreover, in a unique way'.

Consider this identity for an arbitrary finite case, i.e. cut off the multiplication at the lefthand side after the first four factors. The quantities x^k ($k = 0, 1, 2, \ldots$) at the righthand side of this identity can be obtained by multiplying out the braces at the lefthand side. Since the coefficient of each x^k ($k = 0, 1, 2, \ldots$) at the righthand side is 1, this means that each x^k at the righthand side can be obtained in a unique way by multiplying out the braces at the lefthand side. So, in fact each k can be represented as the sum of certain powers of 2 and, moreover, in a unique way.

In turn, Euler concludes from the identity

$$(1 + x + x^{-1})(1 + x^3 + x^{-3})(1 + x^9 + x^{-9}) \cdots = 1 + x + x^{-1} + x^2 + x^{-2} + x^3 + x^{-3} + \ldots$$

'that all [integral] numbers can be obtained from terms of the triple geometrical progression by addition and subtraction, and moreover for each there is a unique manner'.

L. Euler also mentioned a practical side of the problem: 'It is known from the practice of weighing that if we have weights of $1, 2, 4, 8, 16, 32$, etc., pounds, then we can balance all loads with them, if we do not require parts of a pound to be balanced. Moreover, it is known from the practice of weighing that with an even smaller number of weights, growing in triple (i.e. with multiplier 3) geometrical progression, more precisely, of $1, 3, 9, 27, 81$, etc. pounds, we can balance all loads, if we do not require parts to be balanced. However, if it is required to balance in this way we have to place weights not on a single pan but on both pans'.

These facts were successfully used for a long time. Already the Ancient Egyptians used the first of them for multiplying numbers, and Luca Paccioli had zealously struggled with this manner of multiplication. In general, measuring practice has brought about various measuring scales, among which those proportional to 2 or 3 are most widespread. However, among such scales we also encounter such that are, at first glance, not natural, e.g. the Fibonacci sequence.

The very process of weighing has been accompanied at times by special regulations:

'Lifter (outdated) — a trading tax, collected when weighing commodities in certain cities. The naming arose because of the lifting of commodities on weighing instruments. The lifting tax was only due for heavy commodities sold or bought in large quantities, by both the buyer and the seller. The amount due to customs varied from one-half to five coins (Encyclopaedic Dictionary of Brokhaus and Efron).'

Euler took most actively part in the practice of weighing; so, he computed the amount of deviation of the beam of hand balances from being horizontal. In Russia, the 1797 law on measurements and balances prescribed to prepare balance weights of 1 and 2 pud, $1, 3, 9, 27$ pounds, and $1, 3, 9, 27$, and 81 zolotniks.[1] Thus, by and large a triple scale was specified. This testifies of the advanced nature of the measuring thought in Russia. Despite the large spread of special instructions to use such weights, this scale did not last long: one had already 'catched' an unknown integer scale of weighing by the triple scale for two-pan weighing. Already then the factor of computing in real time turned out to be practically significant.

The great Russian chemist D.I. Mendeleev (1834–1907) headed in 1893 the Weights and Measure Office, and remained there as head until his death. He gave theoretical views on the construction of balances, as well as practical conclusions from them, in his work 'On the expansion of gases'. Together with his companions and followers he gave intense attention also to a problem of 'no little importance—the establishment of a rational number of balance weights'. The problem on weight is nowadays called the *Bachet–Mendeleev problem*.

It can be imagined that after the reasoning of Euler the problem was exhaustedly treated—to balance any integral load not heavier than k it suffices to take a smallest necessary amount of weights occurring in the double or triple geometrical progression. So, in the case of p-pan weighing (i.e. weighing for which the weights

---

[1] A zolotnik is 1/96 of a (Russian) pound.

can be put in p pans, $p = 1, 2$) the number of these weights will be a quantity of magnitude $\log_{p+1}(pk + 1)$, since the sum

$$1 + (p+1) + (p+1)^2 + \cdots + (p+1)^{t-1} = \frac{(p+1)^t - 1}{p}$$

must be at least k. For example, if $k = 13$, then for single-pan weighing we can use the weights $(1, 2, 4, 8)$, and for two-pan weighing the weights $(1, 3, 9)$. However, practice turns out to be wider than theoretical conclusions: there are more economical systems of weights. So, to balance a load not heavier than 13 with single-pan weighing we can use other systems of weights, e.g. $(1, 2, 3, 7)$ or $(1, 2, 4, 6)$; they are more economical in the sense that their total sum equals 13, while $1 + 2 + 4 + 8 = 15 > 13$. Why would we use systems of weights whose total sum exceeds the weight of the load to be weighted? Therefore the main question can be stated as

compute the least number of weights, $t_p(k)$, for p-pan weighing, and describe a system of weights, of total sum k, making it possible to weigh any integer load not heavier than k.

The present state of answering this question is given in the following Theorem.

THEOREM 2.5. *Suppose all parameters occurring in this Theorem are integers. Let $[x]$ be the integer part of a number x and $]x[= [-x]$.*

1. *For p-pan weighing, a system of weights $k_1 \geq \cdots \geq k_t$ of total sum k makes it possible to exactly weigh any integer load not heavier than k if and only if*

$$k_i \leq 1 + p \sum_{j=i+1}^{t} k_j, \qquad i = 1, \ldots, t.$$

2. *For p-pan weighing, the number $t_p(k)$, i.e. the least possible number of weights necessary to weigh any integer load not heavier than k, is given by*

$$t_p(k) = \left] \log_{p+1}(pk + 1) \right[.$$

3. *For p-pan weighing, the smallest possible system of weights making it possible to weigh any integer load not heavier than k is recurrently given by*

$$k_i = \left[\frac{pk + 1 - p \sum_{j=1}^{i-1} k_j}{p + 1} \right], \qquad i = 1, \ldots, t = t_p(k);$$

for it we always have

$$k = \sum_{j=1}^{t} k_j.$$

We will demonstrate this assertions of Theorem by a concrete example. Let $k = 40$ and $p = 1$, i.e. we look for the smallest system of weights in single pan weighing. According to the second assertion of the Theorem, $t_1(40) =]\log_2(41)[= 6$. By the third part of the Theorem we can recurrently obtain the weights: $k_1 = [41/2] = 20$, $k_2 = [21/2] = 10$, $k_3 = [11/2] = 5$, $k_4 = [6/2] = 3$, $k_5 = [3/2] = 1$, $k_6 = [2/2] = 1$. These weights, $(20, 10, 5, 3, 1, 1)$, have sum 40. Can we use other systems of weights, e.g. $(19, 10, 6, 3, 1, 1)$? The first part of the Theorem gives a positive answer to this, since it implies that, in particular, $k_i \leq (1 + p)^{t-i}$, $i = 1, \ldots, t$. The fact that $t_2(40) =]\log_3(81)[= \log_3(81) = 4$ gives support to the assumption that we can allow

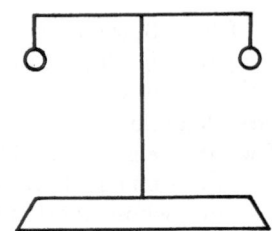

FIGURE 2.4. A drawing by M.V. Lomonosov

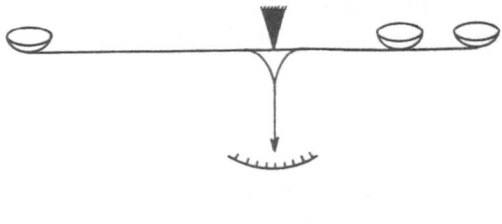

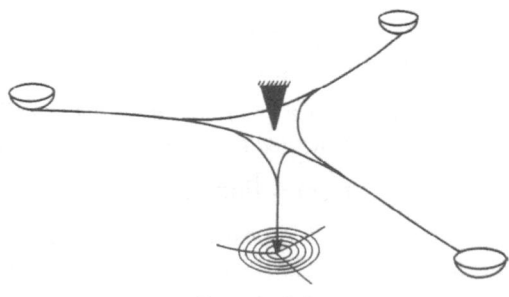

FIGURE 2.5

for two-pan weighing in the original statement of the problem, although Bachet only analyzed single pan weighing.

Engineering problem. It is never asserted in the Theorem that p should only take the values 1, 2. All formulas remain valid also for larger p. Try, 'with the point of a pen', to extract from these formulas balances constructed such that all formulas of the Theorem are realized for each natural number p.

Euler's words that 'questions of this nature clearly exceed the means of analysis without these means' remains true. Therefore the approach expounded above is directly related to the initial problem statements and gives the required information. At the basis of the approach there are extremal results, i.e. results describing only limiting admissible possibilities, and this turns out to be sufficient. The solution of extremal problems leads to the deep structural analysis of a real phenomenon, gives a characterization and concrete realization of it, and, finally, describes general laws or an extremal law.

As already noted, the relation between packability of partitions and weighing is based on the fact that the system of weights $(k_1, \ldots, k_t) \vdash k$ makes it possible to weigh, with single pan weighing, a load $v \le k$ if and only if $(k_1, \ldots, k_t) \subset (v, k - v)$.

Two-pan weighing can be reduced to single pan weighing: a system of weights $(k_1, \ldots, k_t) \vdash k$ makes it possible to weigh, with two-pan weighing, a load $v \le k$ if and only if $(k_1^2, \ldots, k_t^2) \subset (k - v, k + v)$.

In fact, the equality $v = \sum_{j=1}^{t} \epsilon_j k_j$, where $\epsilon_j \in \{0, 1, -1\}$, is equivalent to the equality $k + v = \sum_{j=1}^{t} \eta_j k_j$, where $\eta_j = (\epsilon_j + 1) \in \{0, 1, 2\}$.

Therefore, for p-pan weighing, a system of weights $(k_1, \ldots, k_t) \vdash k$ makes it possible to exactly weigh any integer load not heavier than k if and only if the partition $(k_1^p, \ldots, k_t^p)$ can be packed into all partitions of pk of the form $(k - v, (p - 1)k + v)$, $v = 0, \ldots, k$.

By the principle of complete packability, it is now obvious that a partition $(k_1^p, \ldots, k_t^p)$ can be packed into all partitions of pk of the form $(k - v, (p - 1)k + v)$, $v = 0, \ldots, k$, if and only if $pk \ge n(k_1^p, \ldots, k_t^p; 2)$, which, in turn, by (2.1) is equivalent to the system of inequalities

$$k_i \le 1 + p \sum_{j=i+1}^{t} k_j \le (p + 1)^{t-i}, \qquad i = 1, \ldots, t.$$

Consequently, the least number of weights, $t_p(k)$, needed for such p-pan weighing is the smallest number t such that $pk = m_p(k, t, 2)$, where

$$m_p(k, t, r) = \min_{(k_1, \ldots, k_t) \vdash k} n(k_1^p, \ldots, k_t^p; r).$$

Since

$$m_p(pk, pt, 2) \ge \mu_p(pk, pt, 2) = p\mu \left(k, t, 1 + \tfrac{1}{p} \right),$$

we have $k \ge \mu(k, t, 1 + \tfrac{1}{p})$, and thus, by (2.8), $(p + 1)^t \ge pk + 1$. Taking into account the upper bound for t, we find that $t_p(k)$ is given by the formula $t_p(k) = \left\lceil \log_{p+1}(pk + 1) \right\rceil$, and the weights may be given recursively by, e.g.,

$$y_i = \left\lceil \frac{pk + 1 - p\sum_{j=1}^{i-1} y_j}{p + 1} \right\rceil, \qquad i = 1, \ldots, t_p(k).$$

There is also an explicit solution for p-pan weighing with unique weights; more precisely, for the smallest system of weights we may take

$$K = \left(\left\lceil \frac{pk(p+1)^{t-1}}{(p+1)^t - 1} \right\rceil, \ldots, \left\lceil \frac{pk(p+1)^l}{(p+1)^t - 1} \right\rceil, \left\lceil \frac{pk(p+1)^{l-1}}{(p+1)^t - 1} \right\rceil, \ldots, \left\lceil \frac{pk}{(p+1)^t - 1} \right\rceil \right),$$

where

$$l = \sum_{i=1}^{t} \left\{ \frac{pk(p+1)^{t-i}}{(p+1)^t - 1} \right\} \qquad \text{and} \qquad t = t_p(k).$$

Here, $\{x\}$ denotes the fractional part of the number x, i.e. $\{x\} = x - [x]$. For $p > 2$ the meaning of these expressions is extremely simple, and they can be realized by weighing on balances which are special, but given by a natural construction. They are unequal-arm beam balances, having at the shorter arm a pan for weighing loads only, and at the other arm p pans with mutual distances which are all equal to the length of the smaller arm.

Comparing now the recurrently constructed smallest system of weights from part 3 of the Theorem with the above system for $t = t_p(k)$, we see that they can be different: for $k = 40$ already, the last system has the form $(20, 8, 6, 3, 2, 1)$. This means that the problem of giving an explicit representation of all smallest systems of weights is yet still to be solved.

Already when taking a fleeting glance at the above said there naturally arises the following question: For what weighing procedure can we know the exact values of $m(k, t, r)$ for $r > 2$? In other words, the solution of what engineering problem would give us a computation of $m(k, t, r)$ in full generality? Again, the answer can be seen from the following basic facts. The system of identities described above not only shows the packability $(k_1, \ldots, k_t) \subset (v, k - v)$, but also gives an analytic description of *two* simultaneous weighings of the loads v and $k - v$ with *two independent* balance weights. This leads to the phenomenon of simultaneous (or parallel) weighing using different balance weight and the same system of weights (i.e. the weights taking part on one balance weight do not participate on the other).

One of two worthwhile sellers has gone away and has given his weights to the remaining one, who must simultaneously serve the buyers: This is an illustration for parallel weighing on two balance weights. And so one can readily see the generality of the notion of packability of partitions brought about by the new real situation.

While for single pan weighing and single weighing the sequence

$$1, 2, 4, 8, 16, 32, \ldots$$

turns out to be most efficient, for two independent weighings its analog is the already somewhat more different sequence

$$1, 1, 2, 3, 4, 6, 9, 14, \ldots.$$

At first glance the construction law of this sequence is not clear. However, it can be clearly imagined; we only have to describe the conditions of independent weighing in terms of packability of partitions. Similar to the above we have

- for single pan weighing with $r - 1$ independent balances, weighing loads $(v_1, \ldots, v_{r-1}) \vdash v \leq k$ by weights $(k_1, \ldots, k_t) \vdash k$ can be realized if and only if there is a packing $(k_1, \ldots, k_t) \subseteq (v_1, \ldots, v_{r-1}, k - v)$.
- for single pan weighing with $r - 1$ independent balances, weights $(k_1, \ldots, k_t) \vdash k$ make it possible to simultaneously and exactly weigh any $r - 1$ integer loads of total weight at most k if and only if the partition $(k_1, \ldots, k_t)$ can be packed into every partition of k with at most r parts.

By the principle of complete packability, the latter holds if and only if $k \geq n(k_1, \ldots, k_t; r)$, or

$$k_1 + \cdots + k_t \geq \max_i \left\{ \sum_{j=1}^{i} k_j + (k_i - 1)(r - 1) \right\},$$

or

$$\sum_{j=i+1}^{i} k_j \geq (k_i - 1)(r - 1),$$

or

$$1 + \frac{1}{r-1} \sum_{j=i+1}^{i} k_j \geq k_i.$$

An extremal solution of the latter system of inequalities is given by the sequence $1, 1, 2, 3, 4, 6, 9, 14, \ldots$.

All extremal characteristics for single pan parallel weighing can thus be computed from the extremal results obtained above.

It is not without interest to ask whether we can obtain this strange sequence for parallel weighing with two balances also by the method of generating functions. Or do we here encounter the limit of 'the strength of analysis' from the above-given quotation of Euler?However, the generating function for single pan two-dimensional weighing can be given sufficiently simply:

$$(1 + x + y)^2 (1 + x^2 + y^2)(1 + x^3 + y^3)(1 + x^6 + y^6)(1 + x^9 + y^9) \ldots,$$

or, in the general case,

$$\prod_{i=1}^{\infty} (1 + x^{\alpha_i} + y^{\alpha_i}).$$

Similar to the second equation, Euler writes down the general form of the generating function for two-pan two-dimensional weighing (i.e. with two independent balances):

$$\prod_{i=1}^{\infty} (1 + x^{\alpha_i} + x^{-\alpha_i} + y^{\alpha_i} + y^{-\alpha_i}).$$

However, an indication of the rapidly growing sequence $\{\alpha_i\}$ in this general form has up till now been difficult.

It is clear that such a model of weighing is equivalent to the realization of all outcomes in an allocation scheme. In particular, this implies that for fixed r and n, tending to infinity, the smallest possible number of groups that can realize all outcomes in the scheme of allocating n indistinguishable particles over r indistinguishable boxes behaves asymptotically as

$$\frac{\ln(n + r - 1)}{\ln(r) - \ln(r - 1)},$$

with rest term (which never exceeds zero) of order

$$\frac{\ln(r - 1)}{\ln(r) - \ln(r - 1)}.$$

This system with minimal number of groups is given by the vector Y of Corollary 2.4.

If we allow two-pan weighings on all $r - 1$ balances, and $(k_1, \ldots, k_t) \vdash k$ realizes the simultaneous weighing, then $rk \geq n(k_1^r, \ldots, k_t^r; r)$.

In fact, in two-pan weighing on $r-1$ balances, if a system of weights $(k_1, \ldots, k_t) \vdash k$ simultaneously weights $r - 1$ loads $v_1, \ldots, v_{r-1}$ of total weight at most k, then the system of inequalities

$$v_j = \sum_{i=1}^{t} \epsilon_{i_j} k_i, \qquad \text{where } j = 1, \ldots, r-1; \ \epsilon_{i_j} = 0, 1, -1,$$

$$\sum_{i=1}^{t} |\epsilon_{i_j}| \leq 1, \qquad i = 1, \ldots, t,$$

holds. Clearly, this system is equivalent to the system of inequalities

$$k - v_j = \sum_{i=1}^{t} \eta_{i_j} k_i, \qquad \text{where } j = 1, \ldots, r-1; \ \eta_{i_j} = 1 - \epsilon_{i_j} = 0, 1, 2,$$

$$\sum_{i=1}^{t} |1 - \eta_{i_j}| \leq 1, \qquad i = 1, \ldots, t.$$

Thus, by Dirichlet's principle, $\sum_{i=1}^{t} \eta_{i_j} \leq r$ $(i = 1, \ldots, t)$. Therefore

$$(k_1^r, \ldots, k_t^r) \subseteq (k - v_1, \ldots, k - v_{r-1}, k + v_1 + \cdots + v_{r-1}),$$

i.e. if the system of weights $(k_1, \ldots, k_t) \vdash k$ provides simultaneous weighing of any $r - 1$ integer loads of total weight at most k on $r - 1$ balances, then the partition $(k_1^r, \ldots, k_t^r)$ can be packed into all partitions of rk with r parts for which the $(r-1)$st part does not exceed k. By Theorem 2.3 this is equivalent to the inequality

$$rk \geq n(k_1^r, \ldots, k_t^r; k^{r-1}).$$

But since

$$n(k_1^r, \ldots, k_t^r; k^{r-1}) = n(k_1^r, \ldots, k_t^r; r),$$

this is also equivalent to the inequality

$$rk \geq n(k_1^r, \ldots, k_t^r; r).$$

Thus, if a system of weights $(k_1, \ldots, k_t) \vdash k$ provides simultaneous two-pan weighing of any $r - 1$ integer loads of total weight at most k on $r - 1$ balances, then

$$k_i \leq 1 + r \sum_{j=i+1}^{t} \frac{k_j}{r-1}, \qquad i = 1, \ldots, t,$$

and for the smallest amount, t, of such weights we have the bound

$$t \geq \log_{(2r-1)/(r-1)} \frac{rk + r - 1}{r - 1},$$

since $n(k_1^r, \ldots, k_t^r; r) = rn(k_1^r, \ldots, k_t^r 2 - 1/r)$ and, by Proposition 2.1,

$$m\left(k, t, 1 - \tfrac{1}{r}\right) = \left] \frac{k(2r-1)^t}{(2r-1)^t - (r-1)^t} - \frac{r-1}{r} \right[.$$

In the case of simultaneous two-pan weighing on two balances the system of weights $\{]2^{i-1}[\}_{i=0,1,\ldots}$ is suitable.

Of course, neither Lucas Paccioli, nor Bachet de Méziriac could have imagined that the weighing problem is in itself a computing technique–for fast raising to a power and with computer memory fragmentation.

In the process of functioning of the memory of a computing device there appears a separation between occupied and free parts: fragmentation. If it is then also necessary to enter new information into the computer's memory, e.g. programs and array data, requiring amounts of memory $k_1, \ldots, k_t$, then there arises a natural question: Can the latter be distributed over the fragments of free memory $n_1, \ldots, n_r$? According to the principle of complete packability this can be done of $\sum_{i=1}^{r} n_i \geq n(k_1, \ldots, k_t; r)$.

Weighing makes its appearance into computing techniques, and even in explicit form. The arithmetical balances of Cassini, described in the 'Collection on Devices of the (Paris) Academy of Sciences' (up to 1699), are well known, as is the rising bridge constructed according to the system of general Poncelet in the fortification 'Mont Valerien' near Paris.

On the basis of weighing the numerist-diagrammetrist V.S. Kozlova developed a mechanical image of a computer with screen. On this invention, E. Lucas published his 1890 lecture at the Paresian National Museum of Arts and Trade, where, by the way, one of the two arithmometers for continuous movement of P.L. Chebyshev is kept. The general conclusion of E. Lucas regarding the apparatus of V.S. Kozlova was as follows: 'We think that this apparatus can be utilized, and it will be used in various forms, convenient to some or other requirements of experimenters. The present model of the diagrammeter is only the contemporary hull of the genial idea of Mister Kozlova. I also think that it would be more convenient to replace the hand balances by spring balances. On this apparatus we can obtain the formulas of Simpson, Poncelet, and general Parmantet, and, in general, all quadrature formulas'.

Mendeleev noted that 'in the nature of measurement and balances there are principal instruments of knowledge', that is why all 'notation' scales are important, i.e. allowing one to precisely measure or determine a large amount of unknowns.

A sequence $k_1 \leq k_2 \leq \ldots$ and a natural number p are called a *scale of notation* if each natural number n can be written in the form $n = \sum_{i \geq 1} \epsilon_i k_i$ with $\epsilon_i \in \{0, 1, \ldots, p\}$. Verify that $(\{k_i\}, p)$ is a scale of notation if and only if $k_i \leq 1 + p \sum_{j=1}^{i-1} k_j$, $i = 1, 2, \ldots$.

Try to analyze the Fibonacci sequence as a scale of notation. In solving this question some unexpected surprises will await you.

The above-mentioned Fibonacci representation of numbers by Arab ciphers and zero follows from this too. For the notation of numbers one also uses the well-known Roman ciphers. A distinguished economy of the Roman system is: to write down the first 39 numbers one uses only the 3 ciphers I, V, and X; and for 40 a fourth one: L. This leads to the thought that the Roman system can be characterized by two-pan weighing Try to do this on your own.

These examples show that the results on weighing obtained above can be transferred to various real phenomena related with measuring collections. According to Euler, 'a constant collection is a definite collection, always preserving one and the same value. Such collections are numbers, because they preserve one and the same, once-received value'. Therefore one measures and computes collections by numbers, of which historically the first was the notion of natural number. 'All numbers as we know consist of a certain amount of the unit'; following Euclid, this was the definition given by Diophantus of Alexandrië in his 'Arithmetic'. It can be traced back to Plato and Aristoteles. The Ancient Greeks ascribed the invention of numbers to the legendary Prometeus, son of Femida, goddess of justice, ordinarily depicted with a balance.

We end this Section on the 'everyday phenomenon' of weighing with the following 'dialog':

> '... with the course of time science grows up, and there is the need to combine all that was separate before ... there is the need to change the anatomical study of a subject by a physiological'. (S.M. Solov'ev)

> 'But there has to flow yet quite some water in the Rhine before schools, finally, discover that mathematics can be a humanitarian science and that students can understand Euler equally well as they do Plato and Goethe.' (Andreas Speizer)

8. Lattice structures in the refinement order of number partitions[1]

As noted before, there are several partial orders on the set of partitions: *lexicographic, power set, refinement, partial sums, concatenation*. The *refinement order* on number partitions is perhaps the most natural, although little is known about it. The few authors who have investigated this order structure, have insisted that its 'non-lattice' character is the primary drawback to progress (cf. [198], [151]), for, among other consequences, the computation of its Möbius function becomes unwieldy. Among the few positive results, the authors of [127] have enumerated the number of maximal chains in the refinement order. These observations notwithstanding, our aim in this Section is to show that the refinement order actually carries much lattice structure, even if, as a whole, it is not itself a lattice.

The *refinement order* on the set $P(n)$ of all number partitions of n is the transitive closure of the following successor relation: $m_1 m_2 \ldots m_p \prec m_1' m_2' \ldots m_{p-1}'$, where the set of parts m_i' coincides with that of the parts m_i except for one part $m_i' = m_j + m_k$ for certain $j, k \in \{1, 2, \ldots, p\}$. We illustrate the upward drawing of $P(6)$ in Figure 2.6; notice that it is not a lattice.

Closely related is the set $\Pi(n)$ of all *set partitions* $\{S_i : i \in I\}$ of an n-element set $\{a_1, \ldots, a_n\}$. The S_i's are the *blocks* of the set partition; the *type* of the set partition is the nondecreasing sequence of positive integers corresponding to the sizes of the blocks arranged in nondecreasing order. The set partitions are ordered by: $\{S_i : i \in I\} \leq \{S_j' : j \in J\}$ if, for each $i \in I$, there is a $j \in J$ such that $S_i \subseteq S_j'$. This (refinement) order turns $\Pi(n)$ into a lattice, the much studied *set partition lattice*. Much is known about its algebraic structure (cf. [201]), its Möbius function is known [200], and like the refinement order for number partitions, a precise formula is known for the number of maximal chains between the top and the bottom. $\Pi(4)$ is illustrated in Figure 2.7.

There are at least two other interesting links between the refinement order on number partitions and the partition lattice.

---

[2] This Section was written by Ivan Rival (Dept. Computer Sc. Univ. of Ottawa, Canada) and B. Stechkin (Steklov Math. Inst. of Moscow, Russia), February 18, 1992.

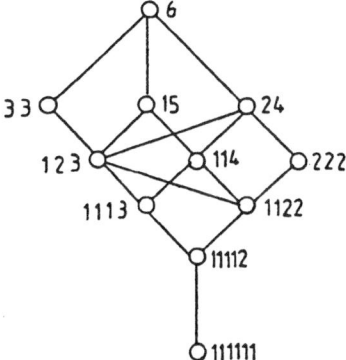

FIGURE 2.6. The refinement order of $P(6)$

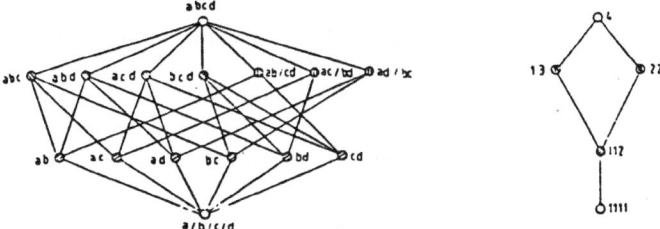

FIGURE 2.7. The refinement order of $\Pi(4)$ on the set of all set partitions of a 4-element set and the corresponding refinement order of $P(4)$ on the collection of its set partition types.

In the first place it is obvious that the types of set partitions are themselves number partitions. We consider an order on the collection of set partition types: $m_1 m_2 \cdots \leq m'_1 m'_2 \ldots$ if there exist corresponding comparable set partitions $\{S_i: i = 1, 2, \ldots\} \leq \{S'_j: j = 1, 2, \ldots\}$ in $\Pi(n)$. (Equivalently, this is the order induced by $\Pi(n)$ on the orbits of its automorphisms.) This natural relation between $\Pi(n)$ and $P(n)$ has been noted by several authors (cf. [151], [79]). We illustrate this for $n = 4$ in Figure 2.7. Of course, $P(n)$ is not order embeddable in $\Pi(n)$: although the lengths of the two orders are identical, $P(n)$ is not a lattice, for every $n \geq 5$.

The second link is not as obvious. It is the substance of our first result, a connection between arithmetic and lattice structure. It is inspired by the simple observation that number partitions with increasing parts sufficiently far apart behave much as set partitions.

THEOREM 2.6.   (i) *For $n \geq p \geq 5$, the up set of all number partitions above $m_1 \ldots m_p$ in $P(n)$ is a lattice if and only if $\sum_{i \in I} m_i \neq \sum_{j \in J} m_j$ for each $I, J \subset \{1, 2, \ldots, p\}$ such that $I \cap J = \emptyset$ and $|I \cup J| < p$.*
  (ii) *Moreover, in this case, the up set above $m_1 \ldots m_p$ in the refinement order of $P(n)$ is isomorphic to the set partition lattice $\Pi(p)$ on a p-element set.*
 (iii) *$\Pi(\log n)$ is the largest set partition lattice isomorphic to an up set in the refinement order of $P(n)$.*

Thus, the number partitions above 1234 in $P(10)$ will not form a lattice, although the number partitions above 1248 in $P(15)$, or even above 1247 in $P(14)$, form a lattice. The refinement order on number partitions also contains large down sets with lattice structure.

PROPOSITION 2.2. *There is a number partition with $\lfloor n/3 \rfloor$ parts such that the down set of all number partitions below it, in $P(n)$, is a distributive lattice of length $\lfloor (2/3)n \rfloor$. Moreover, for $n \geq 6$, $\lfloor (2/3)n \rfloor$ is the length of the longest distributive lattice isomorphic to a down set in $P(n)$.*

Up till now, the calculation of the Möbius function for the refinement order of number partitions has remained virtually inaccessible. At the same time its calculation, on distributive lattices and on set partition lattices, is fully understood. It follows, therefore, from Theorem 2.6 and from Proposition 2.2 that, at least on large intervals of the refinement order for number partitions, we have a full understanding of the Möbius function.

So much for the lattice structure of down sets and up sets in $P(n)$.

Although $P(n)$ is itself not a lattice for every $n \geq 5$, there is a 'regularity' to its upward drawings (cf. Figure 2.9) which suggested to us the idea to delete comparabilities to obtain lattice structure. And, indeed, there is a natural way to define such an order.

We define the *concatenation refinement order* or *C-order* $C(n)$ on the set of all number partitions of n as the transitive closure of this successor relation:

$$m_1 \ldots m_p \prec m'_1 \ldots m'_{p-1}$$

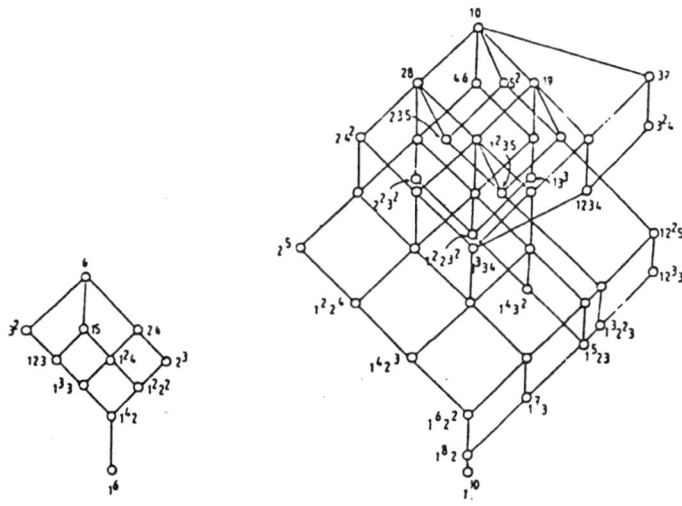

FIGURE 2.8. The C-orders $C(6)$ and $C(10)$

if there is an i such that

$$m'_j = \begin{cases} m_j & \text{if } j < i, \\ m_j + m_{j+1} & \text{if } j = i, \\ m_{j+1} & \text{if } j \geq i = 1. \end{cases}$$

The difference with the refinement order lies in the following two distinctive features of the C-order:

(i) any successor of a number partition is constructed by summing consecutive terms,

(ii) this new summand satisfies the inequality $m_i + m_{i+1} \leq m_{i+2}$.

This inequality condition on the summand suggests a 'Fibonacci' property.

It is convenient to examine the structure of a lattice by identifying its supremum- and infimum-irreducible elements for, as is well known, the subset of these elements determines the lattice. As the elements of the C-order are, in the first place, explicitly described, it is a relatively light matter to describe its irreducibles. For instance, if a number partition $m_1 \ldots m_p$ has parts $m_i, m_j, i \neq j$, such that

$$2m_{i-1} \leq m_i \leq m_{i+1}$$

and

$$2m_{j-1} \leq m_j \leq m_{j+1},$$

then it is supremum irreducible in this C-order.

Although the upward drawing of the refinement order would seem too cluttered to be of any use in 'reading' the order, this C-order is 'weak' enough for its upward drawing to be constructed with little difficulty and to be actually 'read'. For instance, in Figure 2.8, we illustrate the C-order for $n = 10$ and, for convenience, we label only its union-irreducible and intersection-irreducible elements. In Figure 2.9, we illustrate the refinement order of number partitions for $n = 9$. Its elements are traced from their corresponding coordinate positions in Figure 2.8, and its labels

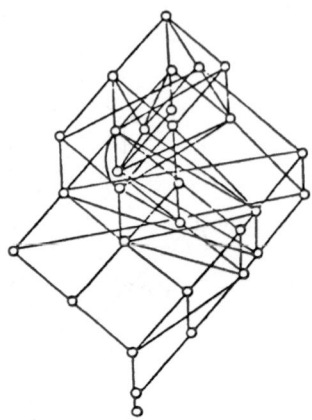

FIGURE 2.9. The upward drawing of $P(9)$

can, thereby, be readily reconstructed from the labels in Figure 2.8 too, as the down set of all number partitions beginning with the part 1.

THEOREM 2.7. *The C-order on the set of all number partitions is a lattice.*

Besides the improved prospects for investigating the order structure of number partitions using these upward drawings, the C-order leads surprisingly to an upper bound on the number $p_r(n)$ of all number partitions of n with r parts. It is well known that

$$p_r(n) \geq \frac{1}{r!}\binom{n-1}{r-1}$$

and, for $r = o(n^{1/3})$, this is (asymptotically) equality [199].

THEOREM 2.8. *For any positive integers $r \leq n$,*

$$p_r(n) \leq \binom{\lfloor \frac{n(r-1)}{r} \rfloor}{r-1}.$$

8.1. Proofs of the results.

PROOF OF THEOREM 2.6. It is easy to verify that $P(n)$ is a lattice for $n = 4$ and for $p = 1, 2, 3$. (There are too few parts in these number partitions to produce much structure at all!) The smallest ordered set with top and bottom which is not a lattice is the 6-element ordered set illustrated in Figure 2.10. If $p = 4$ and all four parts are identical, then the up set above $m_1 m_2 m_3 m_4$ is still a lattice (such a partition has precisely one upper cover). If $p = 4$ but there are just two unequal parts, the up set is still a lattice (its upper cover has at most one number partition with three distinct parts). If $p = 4$ and there are at least three distinct parts, we claim that the assertion of the theorem holds too.

Let $m_1 \ldots m_p \in P(n)$ satisfy $\sum_{i \in I} m_i \neq \sum_{j \in J} m_j$ for every $I, J \subset \{1, \ldots, p\}$ such that $I \cap J = \emptyset$ and $|I \cup J| < p$. As there are no duplicate proper disjoint partial sums, addition of numbers plays the same role as juxtaposition of these numbers as digits. In this sense, the refinement order on the set of all number partitions, above

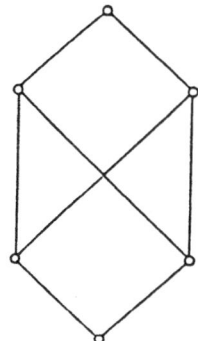

FIGURE 2.10. The smallest non-lattice with top and bottom

$m_1 \ldots m_p$ in $P(n)$, is precisely the same as the refinement order on the set of all set partitions on the p-element set $\{m_1, \ldots, m_p\}$.

Let us suppose now that there exist disjoint subsets I, J of $\{1, \ldots, p\}$, whose union is not all of $\{1, \ldots, p\}$, such that $\sum_{i \in I} m_i = \sum_{j \in J} m_j$. We show, by distinguishing cases, that under these hypotheses $P(n)$ contains a cover-preserving subset isomorphic to the ordered set illustrated in Figure 2.10 (its bottom being a number partition with four parts), whence $P(n)$ itself cannot be a lattice. We first consider the case $|I| > 1$, say $i_0 \in I$,

$$m_{I-i_0} = \sum_{i \in I-\{i_0\}} m_i,$$

$$m_J = \sum_{j \in J} m_j.$$

As $I \cup J \neq \{1, \ldots, p\}$, we can consider also

$$m_K = \sum_{k \in \{1, \ldots, p\} - I \cup J} m_k.$$

Now take the number partitions

$$m_{i_0} + m_k, \quad m_J, \quad m_{I-i_0}$$

and

$$m_J + m_K, \quad m_{i_0}, \quad m_{I-i_0}.$$

They are distinct: $m_J \neq m_J + m_K$ and, since $m_{i_0} + m_{I-i_0} = m_J$, we have $m_j \neq m_{i_0}, m_{I-i_0}$. They are noncomparable in $P(n)$, for they both have three parts. Each of these number partitions covers $m_{i_0}, m_{I-i_0}, m_J, m_K$. Each of these number partitions is covered by the number partitions

$$m_{i_0} + m_{I-i_0} + m_K, \quad m_J \quad \left(= m_{i_0} + m_{I-i_0} = \sum_{i \in I} m_i \right)$$

and

$$m_{I-i_0}, \quad (m_{i_0} + m_K) + m_J \quad (= (m_J + m_K) + m_{i_0}).$$

These are distinct for $m_J > m_{I-i_0}$ and $m_J < m_J + m_K + m_{i_0}$. Since both have two parts, they must be noncomparable. Each, in turn, is covered by top $(= n)$. This

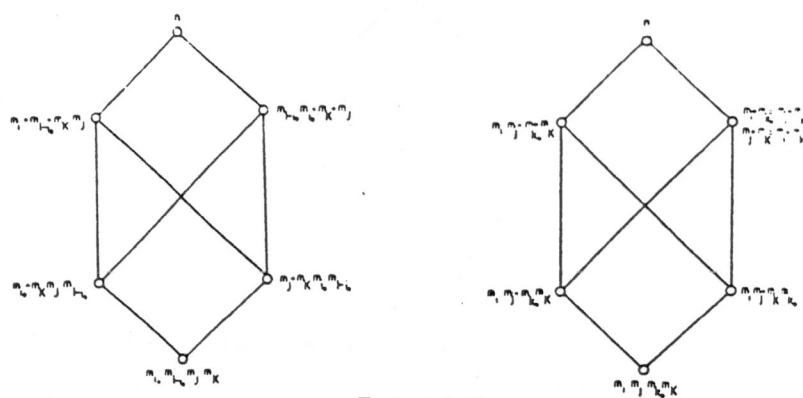

FIGURE 2.11

6-element cover-preserving subset of $P(n)$ is isomorphic to the non-lattice in Figure 2.10, so, in this case, $P(n)$ is not a lattice.

We may suppose now that $m_1 \ldots m_p \in P(n)$ satisfies $\sum_{i \in I} m_i = \sum_{j \in J} m_j$ for some $I, J \subset \{1, \ldots, p\}$ such that $I \cap J = \emptyset$ and $|I \cup J| < p$, *only* in the case $|I| = |J| = 1$. Say $m_i = m_j$, $i \neq j$. Suppose there are two other parts $m_{k_0} \neq m_{k_1}$. Let $m_K = n - (m_i + m_j + m_{k_0})$. Consider the two number partitions

$$m_i, \quad m_j + m_{k_0}, \quad m_K$$

and

$$m_i, \quad m_j + m_K, \quad m_{k_0}.$$

They are distinct:

$$m_j + m_{k_0} \neq m_i, \quad m_j + m_K, \quad m_{k_0}.$$

They are noncomparable in $P(n)$, for they both have three parts. Each of these number partitions covers the number partition m_i, m_j, m_{k_0}, m_K. Each has both

$$m_i, \quad m_j + m_{k_0} + m_K$$

and

$$m_i + m_{k_0} \quad (= m_j + m_{k_0}), \quad m_j + m_K \quad (= m_i + m_K)$$

as upper covers. These upper covers are distinct for $m_j + m_{k_0} + m_K \neq m_j + m_{k_0}, m_j + m_K$ and noncomparable, for each has precisely two parts. These four number partitions, along with bottom $= m_i, m_j, m_{k_0}, m_K$ and top $= n$ form a 6-element cover-preserving lattice in $P(n)$.

If all other parts m_k, $k \neq i, j$, of this number partition $m_1 \ldots m_p$ are identical, $m_k = m_{k'}$ for all $k, k' \in \{1, \ldots, p\} \setminus \{i, j\}$, then take one part m_{k_0} and, for another part, take the partial sum of the remaining terms $m_K = \sum_{r \neq i, j, k_0} m_r$. From the hypothesis that $p \geq 5$ we deduce that $m_{k_0} \neq m_K$ and again we can construct a cover-preserving subset of $P(n)$ which is not a lattice: bottom $= m_i, m_j, m_{k_0}, m_K$, covered by each of the number partitions

$$m_i, \quad m_j + m_{k_0}, \quad m_K$$

and

$$m_i, \quad m_j + m_K, \quad m_{k_0}.$$

In turn, these are both covered by each of

$$m_i, \quad m_j + m_{k_0} + m_K$$

and

$$m_i + m_{k_0} \quad (= m_j + m_{k_0}), \quad m_j + m_K \quad (= m_i + m_K)$$

and top $= n$. This completes the proofs of (i) and (ii).

We turn finally to the proof of (iii). First of all, to see that $P(n)$ does contain an up set isomorphic to $\Pi(\log n)$, just take the number partition consisting of the positive integers $1, 2, 2^2, \ldots, 2^i, \ldots$ which satisfies the conditions set out in (i). (Compare this with the *Bachet problem* [152].)

Let t stand for the lowest level (equivalently, the greatest number of parts) of a number partition of n whose up set in $P(n)$ is isomorphic to the set partition lattice $\Pi(t)$. Notice that $P(n)$ has $\lfloor n/2 \rfloor$ number partitions with two parts, while $\Pi(t)$ has $2^{t-1} - 1$ set partitions (of a t-element set) with two parts. As $\Pi(t)$ is an up set in $P(n)$, we conclude that

$$2^{t-1} \leq \left\lfloor \frac{n}{2} \right\rfloor$$

which completes the proof. $\square$

Another proof of (iii) can be fashioned on this interesting idea. Label the singleton subsets of the hypercube 2^t by the consecutive t parts of the number partition whose up set in $P(n)$ is isomorphic to $\Pi(t)$. Label an arbitrary element of this hypercube by the 'non-repeating' partial sums of the (singleton) parts that lie beneath it in 2^t. According to the condition (i) this is well-defined, and the sum (n) of all the parts cannot exceed the top element of this t-dimensional hypercube.

PROOF OF PROPOSITION 2.2. Let $m_1 \ldots m_p$ be a number partition of n whose down set in $P(n)$ is a lattice. By inspection (cf. Figure 2.6 taking the number of partitions with at least one part equal to 1) $P(5)$ is not a lattice and, as $P(m)$ is isomorphic to a down set of $P(n)$, for every $m \leq n$ it follows that $m_i \leq 4$ for all $i = 1, 2, \ldots, p$. If the number partition has at least one part $m_i = 4$, then every other part must be 1 for, if even one part were 2, then the down set would contain a subset isomorphic to the down set of all number partitions below 24 in $P(6)$(cf. again Figure 2.6); thus, in this case, $p \geq n - 3 \geq \lfloor (2/3)n \rfloor$, and, it is easy to verify that the lattice is distributive. Indeed, it is the linear sum of a chain of length $n - 3$ and a copy of 2^2, the two-dimensional hypercube (all of which is a distributive lattice).

We may suppose then that the largest part is 3. We can verify that the down set of number partitions below any number partition $1^i 2^j 3^k$ is itself a lattice. In fact, such a number partition has precisely two lower covers: $1^{i+2} 2^{j-1} 3^k$ (provided $j \geq 1$) and $1^{i+1} 2^{j+1} 3^{k-1}$ (provided $k \geq 1$). This lattice is the linear sum of the singleton chain, $\lfloor n/3 \rfloor - 1$ copies of 2^2, and another singleton chain (provided $n \equiv 3$ (mod 1)), which, of course, is distributive, and has length $\lfloor (2/3)n \rfloor$. This completes the proof. $\square$

PROOF OF THEOREM 2.7. The C-order $C(n)$ is, of course, contained in the refinement order of $P(n)$, that is, if number partitions $m_1 \ldots m_p \leq m'_1 \ldots m'_q$ in $C(n)$, then $m_1 \ldots m_p \leq m'_1 \ldots m'_q$ in $P(n)$. Moreover, the size of every maximal chain

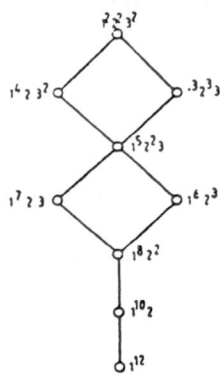

FIGURE 2.12

in $C(n)$ is identical, and identical to the size of every maximal chain in $P(n)$, for between any two comparable number partitions of types $p < q$ there is always a number partition of type $p + 1$, both in $C(n)$ and in $P(n)$. Thus, every covering edge (that is, every edge in the upward drawing) in $C(n)$ is actually a covering edge in $P(n)$.

It is convenient to verify the lattice property using the following criterion: *a finite ordered set with top and bottom is a lattice if and only if every pair of elements covering a common element has supremum* [197]. To this end we observe that $C(n)$ has bottom $= 1^n$ and top $= n$.

Let $m_1 \ldots m_p \in C(n)$. All upper covers have type $p - 1$; indeed, for a distinct pair of upper covers there must be indices $1 \leq i < j \leq p$ such that

$$m_i + m_{i+1} \leq m_{i+2}$$

and

$$m_j + m_{j+1} \leq m_{j+2}.$$

A common upper bound of the number partitions

$$m_1 \ldots m_{i-1} m_i + m_{i+1} \ldots m_p$$

and

$$m_1 \ldots m_{j-1} m_j + m_{j+1} \ldots m_p$$

will have, in the ith place, a part at least $m_i + m_{i+1}$ and, in the jth place, a part at least $m_j + m_{j+1}$. If $i + 1 < j$, then the number partition

$$m_1 \ldots m_i + m_{i+1} \ldots m_j + m_{j+1} \ldots m_p$$

is such an upper bound, of type $p - 2$ and must, therefore, be the supremum of the prescribed pair of upper covers of $m_1 \ldots m_p$.

Let us suppose, then, that $i + 1 = j$. If $p \geq i + 3$ and

$$m_i + m_{i+1} + m_{i+2} \leq m_{i+3}$$

then

$$m_1 \ldots m_i + m_{i+1} + m_{i+2} \ldots m_p$$

is a common upper bound of the number partitions covering $m_1 \ldots m_p$ and, as such, has type $p-2$: it is the required supremum. If $p = i+2$, then $m_1 \ldots m_i + m_{i+1}m_{i+2}$ is the required supremum. Thus, we suppose that $p \geq i+3$ and $m_i + m_{i+1} + m_{i+2} > m_{i+3}$. If $p = i+3$, then $m_1 \ldots m_i + m_{i+1} + m_{i+2} + m_{i+3}$ must be the supremum. Thus, suppose that $p > i+3$.

Let $m'_1 \ldots m'_q$ be the minimal common upper bound of the prescribed pair of upper covers of $m_1 \ldots m_p$. Such an upper bound satisfies $m'_1 = m_1$, $m'_2 = m_2, \ldots, m'_{i-1} = m_{i-1}$, and

$$m'_i \geq m_i + m_{i+1} + m_{i+2}.$$

We give a procedure for the construction of the supremum by considering successively the additive conditions on consecutive terms required for comparability in the C-order. In this way we impose successive conditions on the minimal common upper bound which, in the end, determine it uniquely, thus giving the required supremum. As $m_i + m_{i+1} \leq m_{i+2}$ and $m_j = m_{i+1} + m_{i+2} \leq m_{i+3}$, it follows that

$$m_i + m_{i+1} + m_{i+2} \leq m_{i+3} + m_{i+4}.$$

If $p = i+4$, then $m_1 \ldots m_i + m_{i+1} + m_{i+2}m_{i+3} + m_{i+4}$ is the supremum. Otherwise $p \geq i+5$. Then $m_{i+3} + m_{i+4} \leq m_{i+5}$, and

$$m_1 \ldots m_i + m_{i+1} + m_{i+2}m_{i+3} + m_{i+4} \ldots m_p$$

is the supremum. Otherwise $m_{i+3} + m_{i+4} > m_{i+5}$. Then $p \geq i+6$ and $m_{i+3} + m_{i+4} \leq m_{i+5} + m_{i+6}$.

Continuing in this way we derive necessary additive conditions producing a number partition which, in $C(n)$, is a common upper bound. The procedure terminates by producing the supremum of the two upper covers of $m_1 \ldots m_p$ in $C(n)$ and this completes the proof. $\square$

PROOF OF THEOREM 2.8. Let $t(n,r)$ stand for the smallest integer t for which there is a number partition $m_1 \ldots m_t$ such that, for every number partition $m'_1 \ldots m'_r$ with r parts, $m_1 \ldots m_p \leq m'_1 \ldots m'_r$ in the C-order $C(n)$.

We show first that

$$t(n,r) = \left\lfloor \frac{n(r-1)}{r} \right\rfloor + 1.$$

To see this, observe that the largest part m'_r of $m'_1 \ldots m'_r$ satisfies

$$\left\lceil \frac{n}{r} \right\rceil \leq m'_r \leq n - r + 1.$$

Then the largest part m_t of $m_1 m_2 \ldots m_t$ satisfies

$$m_t \leq \left\lfloor \frac{n}{r} \right\rfloor.$$

Now, $m'_r = \sum_{i=i(m'_r)}^{t} m_i$ and, as long as $m'_r \leq n - r + 1$, there is a number partition with r parts whose largest part has value $m'_r + 1$. It follows that $m'_r = \sum_{i=i(m'_r+1)}^{t} m_i$, where $i(m'_r + 1) < i(m'_r)$. In particular, $i(m'_r + 1) = i(m'_r) - 1$ and $m_{i(m'_r+1)} = 1$. So $m_j = 1$ for all $j \leq i(m'_r + 1)$ As the largest part of a number partition of $P_r(n)$ can be as small as $\lfloor n/r \rfloor$ we conclude that the required number partition $m_1 \ldots m_t$ has the precise form $1^{t-1} \lfloor n/r \rfloor$. We deduce that $t \leq \lfloor n(r-1)/r \rfloor + 1$.

Next we show that there is an order embedding of the subset of all number partitions above $m_1 \ldots m_t$ in $C(n)$ into the $(t-1)$-dimensional hypercube 2^{t-1}. To this end we associate t-tuples of zeros and ones to this up set of number partitions, as follows: $m_1 \ldots m_t$ is associated with the zero sequence $(0, \ldots, 0)$ and a number partition $m'_1 \ldots m'_q \geq m_1 \ldots m_t$ has a 1 in the ith place if in constructing this number partition from $m_1 \ldots m_t$ the sum $m_i + m_{i+1}$ needs to be taken (otherwise 0 in this place). This a one-to-one, order-preserving, and height-preserving map of this up set into 2^{t-1}.

We can now complete the proof. The set $P_r(n)$ of number partitions, each of type r, is an antichain (level) in the up set above $m_1 \ldots m_t$ in $C(n)$, and this antichain, in turn, is associated with a subset of the $(t-r)$th level of the corresponding hypercube 2^{t-1}. In conclusion,

$$p_r(n) \leq \binom{t-1}{t-r} = \binom{t-1}{r-1} \leq \binom{\lfloor \frac{n(r-1)}{r} \rfloor}{r-1}. \qquad \square$$

9. Problems and assertions

In this section we give some well-known results (in the form of problems and assertions requiring proofs) and open problems, which are marked by (?). Try to prove these results and to answer the questions posed.

PROBLEM 2.1. Suppose you stay in a Russian hotel. What is the least amount of banknotes (with values Rbls 1, 3, 5, 10, and 25) you will need to pay a cleaning lady, a nurse, a laundress, and a plumber if your total bill for all services is Rbl 37? How to settle the problem if the ladies will ask for Rbls 3, 5 and 5, respectively?

PROBLEM 2.2. A partition (k^t) can be packed into a partition $(n_1, \ldots, n_r)$ if and only if

$$\sum_{i=1}^{r} \left\lfloor \frac{n_i}{k} \right\rfloor \geq t.$$

Try to consider the dual problem: what partitions $(k_1, \ldots, k_t)$ can be packed into a partition (n^r)?

PROBLEM 2.3. Let $n_a(k, t, r)$ be the largest n for which there is no partition of a number k with t parts that can be packed in any partition of a number n with r parts. Then

$$n_a(k, t, r) = \max\{k-1, k-1-t+r\}.$$

PROBLEM 2.4. If $n_b(k, t, r)$ is the smallest n for which $\forall p \in P_r(n) \; \forall q \in P_t(k) :$ $p \not\subset q$, then $n_b(k, t, r) = \max\{k+1, k+1-t+r\}$.

PROBLEM 2.5. Let $n_c(q_1, \ldots, q_t; r)$ be the smallest n such that no partition of n with r parts can be packed into the partition $(q_1, \ldots, q_t)$, $q_1 \geq \cdots \geq q_t$. Then

$$n_c(q_1, \ldots, q_t; r) = 1 + \sum_{i=1}^{\min(r,t)} q_i.$$

PROBLEM 2.6. If $q \in P_t(k)$ and $n_d(q; r)$ is the largest n for which a partition of q cannot be packed in any partition of n with r parts, then $n_d(q; r) = \max\{k-1, k-1-t+r\}$.

PROBLEM 2.7. If $n_e(k,t,r)$ is the smallest n for which $\forall p \in P_t(n) \, \forall q \in P_t(k) :$ $q \subset p$, then $n_e(k,t,r) = \max\{k, r(k-t)+1\}$.

PROBLEM 2.8. If $n_f(k,t,r)$ is the largest n for which $\forall p \in P_t(n) \, \forall q \in P_t(k) : p \subset q$, then $n_f(k,t,r) = \min\{k,]k/t[+ r - 1\}$.

One of the most fundamental open problems in the theme of extremal number partitions is the following.

PROBLEM 2.9. (?) For given $n_1, \ldots, n_r$, t, compute $k(n_1, \ldots, n_r; t)$, the largest k for which $\forall (k_1, \ldots, k_t) \vdash k$ there holds the packing $(k_1, \ldots, k_t) \subset (n_1, \ldots, n_r)$.

PROBLEM 2.10. A partition $(k_1, \ldots, k_t)$ can be packed into every partition in $P_r(n)$ if and only if the partition $(lk_1, \ldots, lk_t)$ can be packed into every partition in $P_r(nl + (r-1)(l-1))$.

PROBLEM 2.11. A partition $(k_1, \ldots, k_t)$ can be packed into every partition in $P_r(n)$ if and only if the partition $(k_1^l, \ldots, k_t^l)$ can be packed into every partition in $P_{rl-l+1}(nl)$.

PROBLEM 2.12. If a partition $(k_1, \ldots, k_t)$ can be packed into every partition in $P_r(n)$, then the partition $(k_1^l, \ldots, k_t^l)$ can be packed into every partition in $P_r(nl + (r-1)(l-1))$.

PROBLEM 2.13. If, for natural numbers $r, t, n_1, \ldots, n_r, k_1 \geq \cdots \geq k_t$ the inequality

$$\max_{l|k_i} \left\{ \sum_{i=1}^{r} l \left[\frac{n_i}{l}\right] + (r-1)(l-1) \right\} \geq \max_{1 \leq i \leq t} \left\{ \sum_{j=1}^{i} k_j + (r-1)(k_i - 1) \right\}$$

holds, then the partition $(k_1, \ldots, k_t)$ can be packed into the partition $(n_1, \ldots, n_r)$.

PROBLEM 2.14. If $p(k_1, \ldots, k_t; r)$ is the smallest p such that $(k_1, \ldots, k_t) \subset (p^r)$, then

$$p(k_1, \ldots, k_t; r) \leq \left] \frac{n(k_1, \ldots, k_t; r)}{r} \right[.$$

PROBLEM 2.15. If $M(k,t,r)$ is the largest M such that $\forall q \in P_t(k) \, \exists p \in P_r(M) :$ $q \not\subset p$, then for sufficiently large r,

$$M(t,k,r) = \left(\left[\frac{k}{[k/t]}\right] + r - 1 \right) \right] \frac{k}{t} \left[- r.$$

PROBLEM 2.16. If $N(k,t,r)$ is the smallest N such that $\forall p \in P_r(N) \, \exists q \in P_t(k) :$ $p \subset q$, then

$$N(t,k,r) = k + \left(\right] \frac{k}{t} \left[- 1 \right) \max\{0, r-t\}.$$

PROBLEM 2.17. For any allocation of n indistinguishable particles over r indistinguishable boxes there are r groups of particles (with $n/(2r)$ particles in each group) lying completely in the boxes.

PROBLEM 2.18. For any allocation of n indistinguishable particles over r indistinguishable boxes there are t groups of particles (with $[(n+r-1)/(t+r-1)]$ particles in each group) lying completely in the boxes.

PROBLEM 2.19. If $k = n(k_1, \ldots, k_t; r)$, then every allocation of k particles over r boxes can be realized by t groups of particles with k_j particles in group j ($j = 1, \ldots, t$), under the condition that each group lies completely in a single box.

PROBLEM 2.20. If the inequalities $n_1 \geq \cdots \geq n_r$ hold for a partition $(n_1, \ldots, n_r) \vdash n$, then

$$\{(p_1, \ldots, p_r) \in P_r(n) : p_i \leq n_i, \ i = 2, \ldots, r\} \subseteq$$
$$\subseteq \{(p_1, \ldots, p_r) \in P_r(n) : (p_1, \ldots, p_r) \succeq (n_1, \ldots, n_r)\},$$

$$\{(p_1, \ldots, p_r) \in P_r(n) : p_i \leq n_i, \ i = 1, \ldots, r-1\} \subseteq$$
$$\subseteq \{(p_1, \ldots, p_r) \in P_r(n) : (n_1, \ldots, n_r) \succeq (p_1, \ldots, p_r)\}.$$

PROBLEM 2.21. *Counter service problem.* In some city a cooperative confection bar is opened with one long counter. It is known that civilians enter the bar by family (with k_i persons in family i), and that there are t families. Each family is seated at the counter, taking seats in order and not leaving empty seats at both sides simultaneously. Being seated, those who wish may leave, but may come in anew. How many seats must the counter have to prevent queues to form?

If the cooperation includes an administer who seats the guests to his own preferences (not separating families), then how many seats can be spared by him?

PROBLEM 2.22. *Transportation problem.* Suppose we have p warehouses, in each of which m_i units of goods of a single kind are kept ($i = 1, \ldots, p$). Suppose also we have q customers, each of which can use n_j units of goods from the warehouses ($j = 1, \ldots, q$), and suppose, for the sake of being specific, that $m_1 + \cdots + m_p = n_1 + \cdots + n_q = n$. What is the smallest amount of transports necessary to bring all goods from the warehouses to the customers? A single transport is assumed to take a good from one warehouse to a customer.

PROBLEM 2.23. For a natural number n we let $s(n)$ be the number of natural numbers m ($m \geq 2$) for which $m - 1$ divides $[n(m-1)/m]$ without remainder, and we let $d(n)$ be the number of divisors of n. Show that $n - 1$ is a prime if and only if $d(n) = s(n)$. Show that $n - 1$ and $n + 1$ are both primes ('twins') if and only if $2d(n) = s(n) + s(n+1)$.

Hint. Prove and use the equation $d(n-1) + d(n) = s(n) + 2$.

PROBLEM 2.24. Show that the recurrence $k_i = k_{i-1} + k_{i-2}$ ($k_1 = 1$, $k_2 = 2$) and the recurrence

$$k_i = 1 + \max \left\{ \sum_{\substack{j<i \\ j \text{ even}}} k_j, \sum_{\substack{j<i \\ j \text{ odd}}} k_j \right\}, \qquad i = 1, 2, \ldots,$$

define one and the same sequence.

PROBLEM 2.25. Try to verify for the *Frobenius numbers* $s = s(n, n+1, n+p)$ $(p > 1)$ that if $n > p(p-4) + 1$ or $t(p-1) - 2 < n < tp + 1$, $0 < t < p - 3$, then

$$s = n(p-3) + \left[\frac{n}{p}\right](n+1) - 1 + (p-1)\left[\frac{n+1}{p}\right] =$$

$$= n\left(p - 2 + \left[\frac{n}{p}\right]\right) - 1 - \begin{cases} 0, & n \equiv p-1 \pmod{p}, \\ n - p\left[\frac{n}{p}\right], & n \not\equiv p-1 \pmod{p}. \end{cases}$$

For the latest state of the solution of the Frobenius problem we refer to work of I. Kan in the Russian journal 'Discrete Mathematics' (1993).

PROBLEM 2.26. Show that if $n \equiv i \pmod{p}$, $i = 0, \ldots, p-1$, and $f(n) = x_i$, then

$$f(n) = \sum_{i=0}^{p-1} x_i \left(\left[\frac{n+p-i}{p}\right] - \left[\frac{n+p-i-1}{p}\right]\right).$$

(This useful fact was also noted by I. Kan.)

PROBLEM 2.27. Draw the Hasse diagram of P_2, and try to compute its Möbius function.

Extremal problems on graphs and systems of sets

This chapter, which is devoted to extremal problems on graphs and systems of sets, seems at first glance unrelated with the material of the previous chapter, which was concerned with extremal problems on number partitions. However, it turns out that a whole class of problems on extremal properties of graphs (more precisely, the class of problems on local properties of graphs) can in essence be reduced to extremal problems on number partitions, and moreover precisely to such problems as have been considered and solved in Chapter 2.

We introduce the necessary notation. By $S_n = \{a_1, \ldots, a_n\}$, and $S = \{a, \ldots\}$, we denote the indexed, respectively nonindexed, set of vertices of graphs or hypergraphs. A subset of vertices is said to be *independent in the graph* if no pair of vertices in this subset is joined by an edge of this graph.

By $G^2(S_n)$ we denote an arbitrary graph with vertex set S_n, and by G_n^2 a graph on a certain set of n vertices; thus, $G^2(S_n) \subseteq C^2(S_n)$ (for the definition of $C^k(S_n)$ see Chapter 1, §1.4). We also use the following notations for special kinds of graphs:

K_n is the *complete graph* on n vertices, i.e. $K_n = C^2(S_n)$;

$K_{p,q}$ is the *complete bipartite graph* on two vertex sets (the parts) with p and q vertices in each part, respectively; so,, if $S_q \cap S_p = \emptyset$, then $K_{p,q} = C^1(S_p) \cdot C^1(S_q)$;

Z_n is a *star*, i.e. a bipartite graph with one singleton vertex part ($Z_n = K_{1,n}$);

$\bar{G} = C^2(S) \setminus G$ is the graph complementary to the graph G on the vertex set S. For example, the graph complementary to the complete bipartite graph $K_{p,q}$ is the graph $K_p + K_q$ (the graph consisting of two complete subgraphs on two disjoint subsets with p and q vertices, respectively).

For $S \subset S_n$ we denote by $G(S)$ the subgraph of $G(S_n)$ induced (or generated) by the vertex subset S, i.e. consisting of precisely those edges of $G(S_n)$ that join vertices from S; so,

$$G(S) = G(S_n) \cap C^2(S).$$

Two edges of a graph are said to be *independent* if they nonadjacent, i.e. do not have common vertices. A system of pairwise independent edges is called a *pairing*.

$\mathcal{F}_k$ denotes a k-vertex graph with $[k/2]$ independent edges (a pairing);

$\mathcal{F}_k'$ is a pairing with a 'fork', i.e. a k-vertex graph with $]k/2[$ independent (if possible) edges;

C_k is a simple cycle on k vertices;

P_k is a simple path on k vertices.

The *chromatic number* $\chi(G(S_n))$ of an n-vertex graph $G(S_n)$ is defined as the smallest possible amount of colors with which the vertices of S_n can be colored such that vertices joined by an edge in $G(S_n)$ are given different colors. In other words, $\chi(G(S_n))$ is the smallest integer χ for which there is a map $\phi\colon S_n \to [\chi] = \{1,\ldots,\chi\}$ such that for each 'color' $i \in [\chi]$ its complete pre-image under this map, $\phi^{-1}(i) = \{a \in S_n\colon \phi(a) = i\} \subseteq S_n$, is an independent subset of vertices of $G(S_n)$. Hence, for computing the chromatic number of a graph we have to find a partition of the vertex set having least possible rank, under the above-stated conditions on such a partition.

A graph is said to be *connected* if each pair of vertices is joined by a path of graph edges.

A *tree* is a connected graph without cycles.

A *wood* is a graph in which each connectedness component is a tree.

A *hypergraph* on a vertex set S is a subset G of $\mathcal{P}(S)$. The elements of G are subsets $e \subset S$, and are called *hyperedges*. Thus, each hypergraph G is a set $\{e_i\}_{1 \le i \le m}$ of $m = |G|$ hyperedges $e_i \subset S$ on the vertices in S. An *l-graph* is a set $G^l \subset C^l(S)$, so that an ordinary graph is a 2-graph; its elements are simply edges. It is sometimes convenient not to give the vertex set; e.g., a hypergraph F can be specified by its hyperedges $\{A_1,\ldots,A_m\}$, whose order is, in general, immaterial.

For $X \cap S = \emptyset$ the notation $C^k(X) \cdot C^l(S)$ denotes the $(k+l)$-graph on the vertex set $X + S$, of the form $\{e \subset X + S\colon |e \cap X| = k, |e \cap S| = l\}$.

A *system of sets* is a hyperedge of a hypergraph, with the distinction that in a hypergraph all hyperedges are distinct, while the elements of a system of sets may repeat. Thus, a system of sets is a multihypergraph.

By the term *sufficient construct* we understand either a graph, a hypergraph, or a system of sets, satisfying prescribed conditions; a sufficient construct is said to be an *extremal construct* if it is possible in the limit for certain parameters or structural characteristics. For example, the complete n-vertex graph is an extremal construct with largest possible number of edges among all n-vertex graphs.

1. The theorems of Mantel, Turan, and Sperner

Historically the first extremal result on graphs is the following theorem.

MANTEL'S THEOREM. *The largest possible number of edges in an n-vertex graph without triangles is equal to* $[n^2/4]$.

The extremal construct is unique, and is the complete bipartite graph with (if possible) equal parts, i.e. with $[n/2]$ and $]n/2[= n - [n/2]$ vertices in the first and second part, respectively. In turn, an immediate verification shows that $[n/2] \cdot]n/2[= [n^2/4]$.

It is obvious that, according to the box principle, for any three vertices of this graph there is pair of vertices belonging to one part, i.e. a pair not joined by an edge. Hence there is no triangle K_3 on these vertices.

The following is a generalization of this.

TURAN'S THEOREM. *Let $T(n, k, 2)$ be the smallest possible number of edges in an n-vertex graph such that for any k vertices there is at least one edge. Then*

$$T(n, k, 2) = \sum_{i=0}^{k-2} \binom{\left[\frac{n+i}{k-1}\right]}{2}.$$

The extremal construct is unique, and is a system of $k-1$ complete graphs with (if possible) equal number of vertices (with $[(n+i)/(k-1)]$ vertices in the ith complete graph, $i = 0, \ldots, k - 2$).

According to the box principle, among any k vertices there is a pair of vertices belonging to one of these complete subgraphs, and hence a joining edge. It is obvious in what way these two theorems are related. If a graph $G(S_n)$ has the property expressed in Mantel's theorem, i.e. $K_3 \not\subset G(S_n)$, then this is clearly equivalent to the fact that in the complemented graph $\bar{G} = C^2(S_n) \setminus G(S_n)$ there is between any three vertices at least one edge. Hence, $T(n, 3, 2) = \binom{n}{2} - [n^2/4]$.

The similar problem for uniform hypergraphs is open, and is called

TURAN'S PROBLEM. Consider uniform l-graphs $G_n^l \subset C^l(S_n)$ on the vertex set S_n, i.e. hypergraphs with as edges l-element subsets of the vertex set S_n. Let $T(n, k, l)$ be the smallest possible number of l-edges in the n-vertex graph $G_n^l \subseteq C^l(S_n)$ such that

$$\forall S_k \subseteq S_n \, \exists S_l \in G_n^l : \, S_j \subseteq S_k.$$

The problem is to calculate $T(n, k, l)$ for all admissible parameters $n \geq k \geq l \geq 1$.

Besides Turan's theorem, only some particular solutions of this general problem are known, namely:

$$T(n, k, l) = n - k + l \Leftrightarrow n \leq \frac{(n-k)l}{l-1};$$

$$T(n, n - 1, l) = \left]\frac{n}{n-l}\right[;$$

$$T(n, n - 2, n - 3) = \left]\frac{n}{3}\right]\frac{n-1}{2}\left[\right[;$$

$$T(n, k+1, l) = \begin{cases} \left]\frac{n(3l-2)}{l}\right[- 3k, & \frac{l}{l-1} \leq \frac{n}{k} \leq \frac{3l}{3l-4}, \ l \text{ even}, \\ 3n - \left]\frac{k(3l-1)}{l-1}\right[, & \frac{l}{l-1} \leq \frac{n}{k} \leq \frac{3l+1}{3l-3}, \ l \text{ odd}; \end{cases}$$

$$T(n, k+1, 3) = \begin{cases} n - k, & 1 \leq \frac{n}{k} \leq \frac{3}{2}, \\ 3n - 4k, & \frac{3}{2} \leq \frac{n}{k} \leq 2, \\ 4n - 6k, & 2 \leq \frac{n}{k} \leq \frac{9}{4}, \ n \neq \frac{9k-1}{4}, \\ 4n - 6k + 2, & n = \frac{9k+d}{4}, \ d = 1, 2. \end{cases}$$

There is a very attractive conjecture of Turan, stating that $T(2n, 5, 3) = 2\binom{n}{3}$, where this value is realized by the 3-graph $K_n^3 + K_n^3$ consisting of two complete 3-graphs on n vertices each. However, even this particular problem is still open up till now.

More information on Turan numbers can be found in [204]

Altogether, the origin of extremal problems on systems of sets is the following result, now known as *Sperner's theorem*.

SPERNER'S THEOREM. *The largest number of subsets of an n-element set which do not contain each other is equal to*

$$\binom{n}{[n/2]}.$$

An extremal construct realizing this value is, e.g., the set of all $[n/2]$-element subsets of an n-element set, since all subsets of equal cardinality are pairwise not included in each other.

We now dwell on some approaches to the solution of these and related extremal problems. The means most often used is that of two-sided estimation, which we will conveniently demonstrate on a simple concrete example.

EXAMPLE 3.1. How many hyperedges can an n-vertex hypergraph without non-intersecting hyperedges have?

If $G \subset \mathcal{P}(S_n)$ is a sufficient construct and $e \in G$, then, clearly, $(S_n \setminus e) \notin G$, hence G cannot contain more than half the number of all possible hyperedges, i.e.

$$|G| \le \frac{|\mathcal{P}(S_n)|}{2} = \frac{2^n}{2} = 2^{n-1}.$$

On the other hand, for $a \in S_n$ the hypergraph $C(a) \cdot \mathcal{P}(S_n \setminus a)$ has the required property, and has number of edges equal to

$$|C(a) \cdot \mathcal{P}(S_n \setminus a)| = |C(a)| \cdot |\mathcal{P}(S_n \setminus a)| = 1 \cdot 2^{n-1} = 2^{n-1}.$$

At the basis of the majority of estimation means lies the use of various cardinal relations for systems of sets.

LEMMA 3.1. *If a hypergraph F and a system of hypergraphs $\mathcal{W} = \{G, \dots\}$ are such that $\forall G \in \mathcal{W}$: $|G \cap F| \le 1$, then*

$$\sum_{S \in F} \frac{\deg_{\mathcal{W}}(S)}{|\mathcal{W}|} \le 1,$$

where $\deg_{\mathcal{W}}(S) = |\{G \in \mathcal{W} : S \in G\}|$.

PROOF. Recall that for a hypergraph G, vertex set A, and nonnegative integer q the valency is defined as the number $v(A, q, G) = |\{e \in G : |A \cap e| = q\}|$. If G and F are two systems of sets, then

$$\sum_{A \in F} v(A, q, G) = \sum_{B \in G} v(B, q, F).$$

In fact,

$$\sum_{A \in F} v(A, q, G) = \sum_{A \in G} |\{B \in G : |A \cap B| = q\}| =$$

$$= \sum_{A \in F} \sum_{B \in G} \chi\{|A \cap B| = q\} = \sum_{B \in G} \sum_{A \in F} \chi\{|A \cap B| = q\} =$$

$$= \sum_{B \in G} |\{A \in G : |A \cap B| = q\}| = \sum_{B \in G} v(B, q, F).$$

It is obvious that if $S \subseteq S_n$, then $v(S, q; C^l(S_n)) = \binom{|S|}{q}\binom{n-|S|}{l-q}$, therefore for $G_n^l \subset G^l(S_n)$ the above equation gives

$$\sum_{S_p \subseteq S_n} v(S_p, q; G_n^l) = \binom{l}{q}\binom{n-l}{p-q}|G_n^l|.$$

If G is a hypergraph and S is a vertex set, then

$$\sum_{a \in S} \deg_G(a) = \sum_{e \in G} |e \cap S|,$$

where $\deg_G(a) = |\{e \in G: a \in e\}|$ is the degree of the vertex a in the hypergraph G. Indeed,

$$\sum_{a \in S} \deg_G(a) = \sum_{a \in S} |\{e \in G : a \in e\}| = \sum_{a \in S}\sum_{e \in G} \chi\{a \in e\} =$$

$$= \sum_{e \in G}\sum_{a \in S} \chi\{a \in e\} = \sum_{e \in G} |e \cap S|.$$

Hence, if we now take S to be a hypergraph F, and G a system of hypergraphs $\mathcal{W} = \{G, \dots\}$, respectively, we obtain the identity

$$\sum_{S \in F} \deg_{\mathcal{W}}(S) = \sum_{G \in \mathcal{W}} |G \cap F|,$$

which implies the required inequality since $\forall C \in \mathcal{W}$: $|G \cap F| \leq 1$. $\square$

In particular, this lemma allows us to prove Sperner's theorem. The following holds.

COROLLARY 3.1. *If a hypergraph $F \subset \mathcal{P}(S_n)$ has the property that $\forall A, B \in F$: $A \not\subset B$, then the following inequality holds:*

$$\sum_{A \in F} \frac{|A|!\,(n-|A|)!}{n!} \leq 1.$$

PROOF. It suffices to take in Lemma 3.1 a hypergraph F satisfying $\forall A, B \in F$: $A \not\subset B$, and to take for $\mathcal{W}$ the set of all hypergraphs G each of which can be represented by a complete chain of the form $G = \{\{S_0\}, \{S_1\}, \dots, \{S_n\}\}$; such a hypergraph is a chain if $\{S_0 \subset S_1 \subset \cdots \subset S_n\}$. It is obvious that the number of such chains in $\mathcal{P}(S_n)$ is $n!$, hence $|\mathcal{W}| = n!$. Since each hypergraph G is a chain, we have

$$\forall G \in \mathcal{W} : |G \cap F| \leq 1.$$

Finally, it is easy to verify that $\deg_{\mathcal{W}}(S) = |S|!\,(n - |S|)!$. $\square$

Now Sperner's theorem follows immediately from Corollary 3.1, since

$$|F|\frac{[n/2]!\,(n-[n/2])!}{n!} = \sum_{A \in F} \frac{[n/2]!\,(n-[n/2])!}{n!} \leq \sum_{A \in F}\frac{|A|!\,(n-|A|)!}{n!} \leq 1,$$

and hence

$$|F| \leq \binom{n}{[n/2]}.$$

In a number of cases it is possible to obtain an exact solution, without turning to two-sided estimation. This has been realized for monotone properties of graphs and hypergraphs. A property $\mathcal{A}$ is said to be *monotone* (hereditary) if the fact that a graph G has this property (written as $G \in \mathcal{A}$) implies that every subgraph F of it also has this property, i.e. $\mathcal{A}$ is monotone if and only if $G \in \mathcal{A} \Rightarrow \forall F \subseteq G \colon F \in \mathcal{A}$. It turns out that for certain monotone properties it suffices to compute an exact solution for one concrete value of n, after which we verify the rate of growth of the extremal bound looked for as a function of n: if this function satisfies well-known analytical conditions, then it is also the required extremal bound. The basis for this method is given by the following lemma.

LEMMA 3.2. *Suppose we can compute, for an arbitrary monotone property $\mathcal{A}$, the quantity*

$$f(n, l; \mathcal{A}) = \max_{G_n^l \in \mathcal{A}} |G_n^l|.$$

If there exists an integral-valued function $f(n)$ such that

$$f(n_0) = f(n_0, l; \mathcal{A}), \qquad n_0 \geq l, \tag{3.1}$$

$$0 \leq f(n) \leq f(n, l; \mathcal{A}), \qquad n > n_0, \tag{3.2}$$

$$\frac{f(n-1)}{1 + f(n)} \leq \frac{n-l}{n}, \qquad n > n_0, \tag{3.3}$$

then for all $n \geq n_0$ we have

$$f(n, l; \mathcal{A}) = f(n).$$

PROOF. Since $\mathcal{A}$ is monotone, we have

$$G^l(S_n) \in \mathcal{A} \Rightarrow \forall S_{n-1} \subset S_n : G^l(S_{n-1}) = G^l(S_n) \cap C^l(S_{n-1}) \in \mathcal{A},$$

and hence $\forall S_{n-1} \subset S_n \colon |G^l(S_{n-1})| \leq f(n-1, l, \mathcal{A})$. Since for any l-graph $G^l(S_n)$ we have

$$\sum_{S_{n-1} \subset S_n} |G^l(S_{n-1})| = (n-l) \cdot |G^l(S_n)|,$$

we find that if $G^l(S_n)$ is an extremal construct, then the inequality $f(n, l; \mathcal{A}) \leq [nf(n-1, l; \mathcal{A})/(n-l)]$ holds. We now prove the required assertion by induction with respect to $n \geq n_0$. For $n = 1$ it is true by (3.1). Suppose it is true up to $n-1$, inclusive. We show that it is also true for n. By the induction hypothesis we have

$$f(n) \leq f(n, l, \mathcal{A}) \leq \left[\frac{n \cdot f(n-1, l; \mathcal{A})}{n-l} \right] = \left[\frac{n \cdot f(n-1)}{n-l} \right] \leq f(n).$$

Here, the first inequality follows from (3.2), the second has been proved above, the equality is the result of the induction hypothesis, and the last inequality follows from (3.3). □

We demonstrate the application of this lemma on some concrete examples. The function $f(n) = [n^2/4]$ satisfies (3.3) for all $n \geq 2$. Hence, if the property $\mathcal{A}$ is that the graph does not contain triangles, then the complete bipartite graph with (if possible) equal parts is in fact the extremal construct in Mantel's theorem, since conditions (3.1), (3.2) clearly hold for $n_0 = 2$. Thus, Mantel's theorem follows immediately from Lemma 3.2.

The verification of the threshold value n_0 in the lemma is of utmost importance. To convince ourselves of this, we consider another monotone property. Let $\mathcal{A}$ be the property that the graph does not contain pentagons, i.e. 5-vertex cycles. Put

$$f(n; C_5) = \max_{C_5 \not\subseteq G_n} |G_n|;$$

we show that

$$f(n; C_5) = \begin{cases} \binom{n}{2} & \text{if } 2 \leq n \leq 4, \\ 7 & \text{if } n = 5, \\ \left[\frac{n^2}{4}\right] & \text{if } n \geq 6. \end{cases}$$

For $n \leq 5$ the extremal value can easily be verified directly, and hence, although the function $f(n) = [n^2/4]$ satisfies (3.3) for all $n \geq 2$, we find that the threshold value for this $\mathcal{A}$ is $n_0 = 6$. For $n = 5$ an extremal construct is the graph K_4 completed by one more edge, incident with the fifth vertex; for $n \geq 6$ an extremal construct is the same complete bipartite graph as in Mantel's theorem.

2. Forbidden subgraphs and local properties

Mantel's result may serve as a model example for a large class of extremal problems, whose general form is as follows.

For a fixed graph F we compute the largest possible number of edges in an n-vertex graph not containing F as a subgraph.

It is, however, convenient to consider a somewhat more general statement, which we call the *problem on forbidden subgraphs*.

PROBLEM ON FORBIDDEN SUBGRAPHS. Let $L = \{G, \dots\}$ be a list of fixed 'forbidden' subgraphs. It is required to compute the quantity $f(n; L)$ equal to the largest possible number of edges of an n-vertex graph not containing as a subgraph any graph on the list L.

In the light of this formulation, Turan's result takes the form

$$f(n, K_k) = \binom{n}{2} - T(n, k, 2).$$

Well-known formulas for certain lists of forbidden subgraphs are given in the Problem Section.

The following theorem gives an asymptotic solution of the problem on forbidden subgraphs.

ERDÖS–SHIMONOVICH THEOREM.

$$f(n, L) = \left(1 + \frac{1}{1 - \max_{G \in L} \chi(G)}\right) \frac{n^2}{2} + o(n^2).$$

It must be noted that the main characteristic determining the coefficient at the principal term of the asymptotic expansion in this Theorem is the chromatic number, i.e. a graph characteristic which is difficult to compute.

The Erdös–Shimonovich theorem makes clear that the coefficient at the principal term of the asymptotic expansion is distinct from zero if and only if the list of forbidden subgraphs does not contain two-colorable ones, i.e. graphs colorable with two colors. The case when two-colorable forbidden subgraphs are on this list is termed 'degenerate', and in this case the problem of computing $f(n; L)$ becomes that of estimating the remainder $o(n^2)$. The region of 'most degenerated' cases has been studied.

The equality $f(n; L) = O(n)$ holds if and only if the list of forbidden subgraphs contains either a tree, or a wood (for a finite L).

The equality $f(n; L) = O(1)$ holds if and only if the list of forbidden subgraphs contains pairings and stars.

As an example we show that

$$f(n; K_{r,s}) \leq 0.5 \cdot (s-1)^{1/r} n^{2-1/r} + O(n), \qquad r \leq s, \quad n \to \infty.$$

If $d_1, \ldots, d_n$ are the vertex degrees of a graph G, then the number of stars $K_{r,1}$ in G equals

$$\sum_{i=1}^{n} \binom{d_i}{r}.$$

If G does not contain the complete bipartite graph $K_{r,s}$ as a subgraph, then for each r vertices there can be at most $s - 1$ adjacent, hence the total number of stars $K_{r,1}$ cannot exceed $(s-1)\binom{n}{r}$, and the inequality

$$\sum_{i=1}^{n} \binom{d_i}{r} \leq (s-1)\binom{n}{r}$$

implies the required inequality:

$$2|G| = \sum_{i=1}^{n} d_i \leq (s-1)^{1/r} n^{2-1/r} + O(n).$$

For other 'degenerate' cases see the Problem Section of this Chapter.

The same result of Mantel may serve as a model example of another class of extremal problems, whose general form is as follows.

PROBLEM ON LOCAL PROPERTIES. For a fixed k-vertex graph H_k, compute $m(n; H_k)$, i.e. the smallest possible number of edges in an n-vertex graph G_n each k-vertex subgraph of which contains a subgraph isomorphic to H_k.

The class of problems on local properties is a subclass of the class of problems on forbidden subgraphs; this inclusion gives the obvious equation

$$m(n; H_k) = \binom{n}{2} - f(n; \{G_k : G_k \not\subset \bar{H}_k\}).$$

Let $\mu(n; F_k)$ be the maximal number of edges in an n-vertex graph such that any k-vertex subgraph is imbedded in the k-vertex graph F_k. Clearly,

$$m(n; H_k) + \mu(n; \bar{H}_k) = \binom{n}{2},$$

therefore in the sequel we assume that $F_k = \bar{H}_k$, and we will choose for our convenience one or the other notation.

3. Exact solutions for local properties of graphs

In this Section we give the solution of certain concrete extremal problems on local properties of graphs. We introduce some helpful notation. Let $\Delta(G)$ be the maximal degree of the graph G. Let $Z(a)$ be the set of vertices adjacent to a. Let $t(G)$ be the largest number of independent edges in G.

THEOREM 3.1. *Let H_k be an arbitrary fixed k-vertex graph having vertex degree $k - 1$, and let $F_k = \bar{K}_k$. Let $\mathcal{F}_k$ be the pairing on k vertices. Then a graph G_n has the property that each of its induced k-subgraphs contains H_k, i.e.*

$$\forall S_k \subset S_n : \ G_n(S_k) = C^2(S_k) \cap G_n \supset H_k,$$

if and only if $G_n \supset K_n - F_k$ for $F_k \not\supset \mathcal{F}_k$ and $G_n \supset K_n - F_k$ or $G_n \supset K_n - \mathcal{F}_n$ for $F_k \supset \mathcal{F}_k$.

PROOF. The sufficiency of this assertion is obvious. We prove its necessity. Suppose the graph G_n has the indicated property. We introduce in the graph complementary to it the system $\mathcal{F}_{2t}$ of t independent edges.

The inequality $t > t(F_k)$ can only be fulfilled if $F_k \supset \mathcal{F}_k$, but in this case we must have $\bar{G}_n = \mathcal{F}_{2t} \subset \mathcal{F}_n$, since by adding an edge e to $\mathcal{F}_{2t}$ (e must be adjacent to an edge of $\mathcal{F}_{2t}$, since the latter is maximal) we obtain the graph $\mathcal{F}_{2t} + \{e\}$, in which there are always k vertices S_k such that the proper subgraph $(\mathcal{F}_{2t} + \{e\})(S_k)$ induced by it does not have isolated vertices, the prescribed graph F_k by the condition

$$H_k \supset K_{1,k-1}.$$

Let now $t \leq t(F_k)$. In this case the number of nonisolated vertices of the graph complementary to G_n does not exceed $k - 1$. In fact, choosing in the opposite case for $S_k \subset S_n$ any set of k nonisolated vertices including the $2t$-vertex graph $\mathcal{F}_{2t}$, by the maximality of $\mathcal{F}_{2t}$ we find that there is no proper subgraph $\bar{G}_n(S_k)$ on these k isolated vertices, contradicting the conditions of the theorem. It is obvious that a graph on at most $k - 1$ nonisolated vertices is a subgraph of F_k. □

As a consequence we obtain the solution of a problem on local properties of graphs.

COROLLARY 3.2. *Suppose the k-vertex graph F_k has an isolated vertex. Then*

$$\mu(n; F_k) = \begin{cases} |F_k| & \text{if } F_k \not\supset \mathcal{F}_k, \\ \max([n/2], |F_k|) & \text{if } F_k \supset \mathcal{F}_k. \end{cases}$$

Thus, if in a problem on local properties the graph H_k has a vertex of degree $k - 1$, then the exact solution of this problem is given by Corollary 3.2.

THEOREM 3.2. $m(n; C_k) =]n(n - k + 2)/2[$.

PROOF. It is obvious that for $k = 3$ the complete graph is extremal, while for $k = 4$ so is the complete graph without pairings, hence we assume in the sequel that $k \geq 5$.

If G_n is a sufficient construct, then each of its vertex degrees is at least $n - k + 2$. In fact, if there is a vertex $a \in S_n$ with degree $d(a) \leq n - k + 1$, then there is a vertex subset $S_k \subset S_n$ containing this vertex and such that $d_{G(S_k)}(a) \leq 1$ (here, $d_{G(S_k)}(a)$ is the vertex degree of a in $G(S_k)$). This means that $G(S_k) \supset C_k$. Since the sum of the vertex degrees equals twice the number of its edges, we find that $|G_n| \geq]n(n - k + 2)/2[$.

For the description of extremal constructs on the vertex set S_n we introduce the distance d by the formula

$$d(a_i, a_j) = \min(|i - j|, n - |i - j|).$$

The following inequality holds for this distance: If $1 \leq i_1 < \cdots < i_k \leq n$, then

$$d(a_{i_1}, a_{i_k}) \leq \min \left(\sum_{j=1}^{k-1} d(a_{i_j}, a_{i_{j+1}}), n - \sum_{j=1}^{k-1} d(a_{i_j}, a_{i_{j+1}}) \right). \tag{3.4}$$

The *degree cycle* C_n^t on n vertices is defined as the graph $C_n^t = \{(a_i, a_j): 1 \leq d(a_i, a_j) \leq t\}$, so that $C_n^1 = C_n$. Consequently, in C_n^t every vertex has degree $\min(2t, n - 1)$, and

$$|C_n^t| = \min \left(nt, \binom{n}{2} \right).$$

We show that C_n^t is a sufficient construct for $t = [(n - k + 2)/2]$. Let $S_k = \{a_{i_1}, \ldots, a_{i_k}\} \subset S_n$, where $1 \leq i_1 < \cdots < i_k \leq n$. If $d(a_{i_j}, a_{i_{j+1}}) \leq t$ $(j = 1, \ldots, k-1)$ and $d(a_{i_1}, a_{i_k}) \leq t$, then

$$C_n^t(S_k) \supset C_k = \{(a_{i_1}, a_{i_2}), \ldots, (a_{i_{k-1}}, a_{i_k}), (a_{i_k}, a_{i_1})\}.$$

We now consider the opposite situation. Suppose, for the sake of being specific, that $d(a_{i_j}, a_{i_{j+1}}) \geq t + 1$. Then, for $1 \leq i_j < i_{j+1} \leq i_k$,

$$d(a_{i_j}, a_{i_{j+1}}) \leq]\frac{n - k + 2}{2}[+ l - 2. \tag{3.5}$$

In fact, by (3.4),

$$d(a_{i_1}, a_{i_k}) \leq d(a_{i_1}, a_{i_2}) - \cdots - d(a_{i_{j-1}}, a_{i_j}) - d(a_{i_j}, a_{i_{j+l}}) - d(a_{i_{j+l}}, a_{i_{j+l+1}}) - \cdots$$
$$\cdots - d(a_{i_{k-1}}, a_{i_k}) \leq n - (k - l - 1) - d(a_{i_j}, a_{i_{j+l}}),$$

whence $t + 1 \leq n - k + l + 1 - d(a_{i_j}, a_{i_{j+l}})$. Since the righthand side of (3.5) does not exceed t for $l = 1$, we always find

$$C_n^t(S_k) \supset P_k = \{(a_{i_1}, a_{i_2}), \ldots, (a_{i_{k-1}}, a_{i_k})\}.$$

Thus, if $n - k$ is even, then for k even,

$$C_n^t(S_k) \supset C_k = \{(a_{i_1}, a_{i_3}), (a_{i_3}, a_{i_5}) \ldots, (a_{i_{k-3}}, a_{i_{k-1}}),$$
$$(a_{i_{k-1}}, a_{i_k}), (a_{i_k}, a_{i_{k-2}}), \ldots, (a_{i_4}, a_{i_2}), (a_{i_2}, a_{i_1})\},$$

while for k odd,

$$C_n^t(S_k) \supset C_k = \{(a_{i_1}, a_{i_3}), (a_{i_3}, a_{i_5}) \ldots, (a_{i_{k-2}}, a_{i_k}),$$
$$(a_{i_k}, a_{i_{k-1}}), (a_{i_{k-1}}, a_{i_{k-3}}), \ldots, (a_{i_4}, a_{i_2}), (a_{i_2}, a_{i_1})\}.$$

So, if $n - k$ is even, the degree cycle C_n^t is extremal.

Let now $n - k$ be odd. If $d(a_{i_1}, a_{i_k}) \geq t + 2$, then $t + 2 \leq n - k + 1 - d(a_{i_j}, a_{i_{j+l}})$, hence $d(a_{i_j}, a_{i_{j+l}}) \leq](n - k + 2)/2[+ l - 3$, and for $l = 2$ this quantity does not exceed t. Hence, also in this case $C_n^t(S_k) \supset C_k$. Thus, it remains to consider the

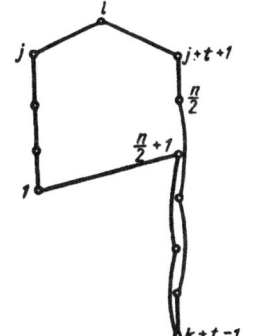

FIGURE 3.1

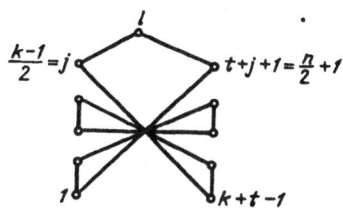

FIGURE 3.2

case $d(a_{i_1}, a_{i_k}) = t+1$. Here, if $d(a_{i_j}, a_{i_{j+2}}) \le t$ $(j = 1, \ldots, k-2)$, then $C_n^t(S_n) \supset C_k$, so we may assume that $d(a_{i_1}, a_{i_k}) = t + 1 = d(a_{i_j}, a_{i_{j+2}})$. But then

$$d(a_{i_1}, a_{i_2}) = \cdots = d(a_{i_{j-1}}, a_{i_j}) = d(a_{i_{j+2}}, a_{i_{j+3}}) = \cdots = d(a_{i_{k-1}}, a_{i_k}) = 1.$$

Therefore, to shorten notation we assume in the sequel that

$$S_k = \{a_1, \ldots, a_j, a_l, a_{j+t+1}, \ldots, a_{k+t-1}\},$$

where $j + 1 \le l \le j + t$, $1 \le j \le [(k-1)/2]$. Now we consider several cases.

Assume that n is even and k is odd. We complete C_n^t by the pairing

$$\mathcal{F} = \{(a_i, a_j) : \ d(a_{i_1}, a_{i_j}) = n/2\},$$

and denote the resulting graph by G. We verify its sufficiency.

Let $t > 1$, $j < (k - 1)/2$. It is obvious that $(a_1, a_{n/2+1}) \in G(S_k)$, and since $j < (k - 1)/2$, we have $t + j + 1 < n/2 + 1 < k + t - 1$. Hence $G(S_k) \supset C_k$, where C_k consists of the path

$$\{(a_{n/2+1}, a_1), (a_1, a_2), \ldots, (a_j, a_l), (a_l, a_{t+j+1}), (a_{t+j+1}, a_{t+j+2}), \ldots, (a_{n/2-1}, a_{n/2})\}$$

and the path spanned on the vertices

$$\{a_{n/2}, a_{n/2+1}, \ldots, a_{k+t-1}\}$$

and beginning in $a_{n/2}$ and ending in $a_{n/2+1}$. (The existence of such a path follows from the condition $t > 1$, cf. Figure 3.1.)

Let $t > 1$, $j = (k - 1)/2$ and j odd. Then $n/2 + 1 = t + j + 1$, and hence $G(S_k) \supset C_k$, with C_k as in Figure 3.2.

Let $t > 1$, $j = (k-1)/2$ and j even. Then either $(a_{j-1}, a_l) \in G(S_k)$ or $(a_l, a_{j+t-2}) \in G(S_k)$. For the sake of begin specific we assume that the first condition holds. But then $G(S_k) \supset C_k$, with C_k as in Figure 3.3.

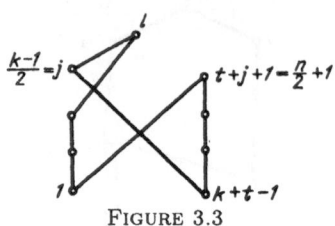

FIGURE 3.3

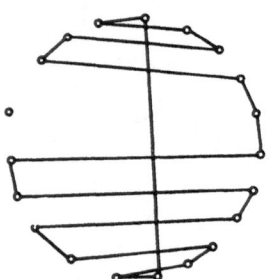

FIGURE 3.4

Let $t = 1$. Then $n = k + 1$, and the graph

$$C_n + \{(a_1, a_{n/2+1}), (a_2, a_n), (a_3, a_{n-1}), \dots, (a_{n/2}, a_{n/2+2})\}$$

is extremal.

In fact, if we remove the vertex a_1, then

$$C_{n-1} = \{(a_2, a_3), (a_3, a_4), \dots, (a_{n-1}, a_n), (a_n, a_2)\},$$

and if we remove a vertex a_j, $j \neq 1, n/2+1$, then C_{n-1} has the shape as in Figure 3.4.

We assume that n is odd and k is even. We complete C_n^t by the pairing with a 'fork'

$$\mathcal{F}' = \{(a_1, a_{(n+1)/2}), (a_i, a_j) : j - i = (n + 1)/2, \ i = 1, \dots, (n - 1)/2\},$$

and denote the resulting graph by G'. We analyze its sufficiency.

We first note that G' does not possess the symmetry of choice of S_k that G has, hence the mastering of the various S_k in G' had better be replaced by the mastering of two distinct pairings with a 'fork': the initial one and the following:

$$\mathcal{F}' = \{(a_{(n-1)/2}, a_n), (a_i, a_j) : j - i = (n - 1)/2, \ i = 1, \dots, (n - 1)/2\}.$$

The equivalence of these masterings is obvious, and their presence makes clear the larger amount of variants than in the case of G. We denote the first variant by a), and the second by b).

a) $t > 1$. Since always $j \leq [(k - 1)/2] = (k - 2)/2 < k$, we have $t + j + 1 < (n + 3)/2 < k + t - 1$, and hence $G'(S_k) \supset C_k$, with C_k as in the case of G for even n, odd k, and $t > 1$, $j < (k - 1)/2$.

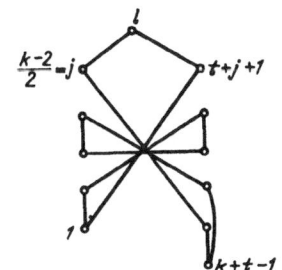

FIGURE 3.5

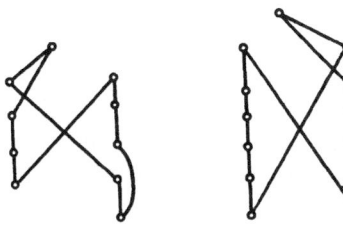

FIGURE 3.6

b) $t > 1$, $j < (k-2)/2$. Since $j < (k-2)/2$, we have $t + j + 1 < (n+1)/2 < k + t - 1$, and hence $G'(S_k) \supset C_k$, as in the previous case.

b) $t > 1$, $j = (k-2)/2$, j odd. Here $G'(S_k) \supset C_k$, with C_k as in Figure 3.5.

b) $t > 1$, $j = (k-2)/2$, j even. Here either $(a_{j-1}, a_l) \in G'$ or $(a_l, a_{t+j+2}) \in G'$. But then $G'(S_k) \supset C_k$, with C_k as in Figure 3.6.

Finally, let $t = 1$. Then $n = k + 1$, and the extremal graph is

$$C_n + \{(a_1, a_{(n+3)/2}), (a_1, a_{(n+1)/2}), (a_2, a_n), (a_3, a_{n-1}), \ldots, (a_{(n-1)/2}, a_{(n+5)/2})\}.$$

In fact, if we delete the vertex a_1, then

$$C_{n-1} = \{(a_2, a_3), (a_3, a_4), \ldots, (a_{n-1}, a_n, a_2)\},$$

and if we delete a vertex a_i, $i \neq 1, (n+3)/2, (n+1)/2$, then C_{n-1} has the shape as in Figure 3.7a), while if a vertex a_i, $i = 1, (n+3)/2, (n+1)/2$, is deleted, then C_{n-1} has the shape as in Figure 3.7b).   □

We note another consequence, concerning the structure of graphs, of the results obtained. We call an n-vertex graph *locally Hamiltonian* if for a natural number k ($3 \leq k \leq n$) every k-vertex proper subgraph of it is Hamiltonian, i.e. is a Hamiltonian cycle C_k. As usual, a *Hamiltonian path* is a simple path passing through all vertices of the graph only once.

COROLLARY 3.3. *In a locally Hamiltonian graph there issues from each vertex a Hamiltonian path.*

PROOF. Assume the contrary: the locally Hamiltonian graph G_n has a vertex a_1 from which no Hamiltonian path issues. Let $P_l = \{(a_1, a_2), \ldots, (a_{l-1}, a_l)\}$ be

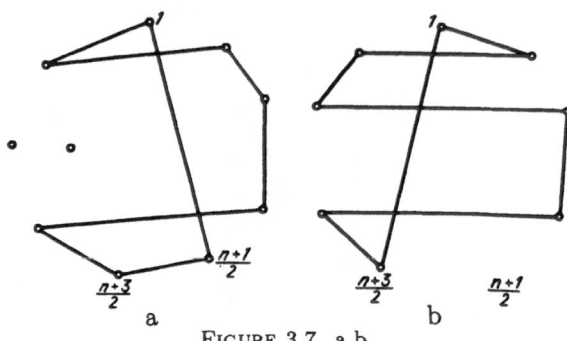

$$\text{FIGURE 3.7. a b}$$

the longest path issuing from a_1. It is obvious that $k \le l \le n - 1$. Put $S_k = \{a_t, \ldots, a_l, \ldots, a_{t+k-1}\}$, where $n - k + 1 > t > l - k + 1, l \ge t \ge 2$ (this can always be realized). Since G_n is locally Hamiltonian, $G_n(S_k) \supset C_k$, hence in $G_n(S_k)$ a Hamiltonian path issues from each vertex, and thus also from a_t. Consider such a path issuing from a_t. In G_n we then have, setting

$$P_t = \{(a_1, a_2), \ldots, (a_{t-1}, a_t)\} \subset P_l,$$

a path P_k starting at a_t and not passing through $\{a_1, \ldots, a_{t-1}\}$ (by construction). Hence there is a path $P_t \cup P_l$ in G_n, issuing from a_1 through $t + k - 1 > l$ vertices, contradicting the maximality of P_l. $\square$

It is remarkable that the formula of Corollary 3.2 remains asymptotically the same, even if the volume of the locally Hamiltonian graph is divided by 2. This is expressed in the following corollary.

COROLLARY 3.4. *If $m(n; \mathcal{F}_k')$ denotes the smallest possible number of edges in an n-vertex graph such that among any k vertices there is a pairing with a 'fork', then*

$$m(n; \mathcal{F}_k') = \left] \frac{n(n - k + 1)}{2} \right[.$$

Considering the result of this corollary, P. Erdös became interested in a condition for the presence of systems of certain independent subgraphs in a local graph, where the systems are taken to be less trivial than a pairing, and in the first instance consist of two independent triangles. The following theorems allow us to find such conditions.

THEOREM 3.3. *For $n \geq 2p + q$,*

$$\mu(n; K_{p,p+q}) = \begin{cases} \binom{n}{2}, & p = 1,\, q = 0, & (3.6) \\[4pt] \left[\frac{n^2}{4}\right], & p = 1 = q, & (3.7) \\[4pt] n - 1, & p = 1,\, q \geq 2, & (3.8) \\[4pt] \max(p^2, n), & p \geq 2,\, q = 0, & (3.9) \\[4pt] 3\left[\frac{n}{2}\right], & p = 2,\, q = 1,\, n = 5, 7, 9, & (3.10) \\[4pt] \left[\frac{3n}{2}\right], & p = 2,\, q = 1,\, n \neq 5, 7, 9, & (3.11) \\[4pt] p^2 + p, & p \geq 3,\, q = 1,\, n = 2p + 1, & (3.12) \\[4pt] \max\left((p+1)^2, 3\left[\frac{n}{2}\right]\right), & p \geq 3,\, q = 1,\, |n \geq 2p + 2, & (3.13) \\[4pt] p^2 + pq, & p \geq 2,\, q \geq 2. & (3.14) \end{cases}$$

PROOF. Let $f(\Delta, t)$ be the maximal number of edges in the graph with maximal degree at most Δ and number of independent edges at most t. Then the estimate $f(\Delta, yt) \leq t(\Delta + 1)$ holds, and the following exact formula is true:

$$f(\Delta, t) = \begin{cases} \Delta t & \text{if } \Delta \geq 2t + 1, \\[4pt] \Delta t + \left[\frac{\Delta}{2}\right]\left[\frac{t}{[(\Delta+1)/2]}\right] & \text{if } \Delta \leq 2t. \end{cases}$$

We show the formulas (3.6)–(3.14), and give the corresponding equation number before proving it.

(3.6). Trivial.

(3.7). The problem is equivalent to computing the maximum number of edges in an n-vertex graph without triangles, and thus equivalent to Mantel's theorem, since the only extremal construct is the complete bipartite graph with (if possible) equal parts.

(3.8). The sufficient construct G_n cannot contain two independent edges, i.e. $|G_n| \leq n - 1$, and this value is realized by the star $K_{1,n-1}$, which is again the unique extremal construct.

(3.9). For $p > 1$ the maximal degree of $K_{p,p+q}$ is $p + q < 2p + q - 1$. Hence $\Delta(G_n) \leq \Delta(K_{p,p+q}) = p + q$, and hence $|G_n| \leq [n(p+q)/2]$. Let G_n be an arbitrary sufficient construct and $a \in S_n$ a vertex of degree $\Delta = \Delta(G_n)$. Since $K_{p,p} \not\supset K_3$, we have $G_n \not\supset K_3$, hence $G_n(\{a\} + Z(a)) = K_{1,\Delta}$. Put $S = S_n \setminus \{a\} \setminus Z(a)$, and consider the induced $(n - \Delta - 1)$-vertex subgraph $G(S) \subset G_n$. If $\Delta \geq 3$, then $t(G(S)) \leq p - \Delta$. In fact, since $K_{p,p} \not\supset K_{1,\Delta} + \mathcal{F}_{2(p-\Delta+1)}$, we have $G(S) \not\supset \mathcal{F}_{2(p-\Delta+1)}$ for $2p \geq 1 + \Delta + 2p - 2\Delta + 2$, or $\Delta \geq 3$. Thus, in this case,

$$|G_n| \leq \Delta^2 + |G(S)| \leq \Delta^2 + f(\Delta, p - \Delta) \leq \Delta^2 + (p - \Delta)(\Delta + 1) =$$
$$= p\Delta + p - \Delta \leq p^2.$$

If $\Delta \leq 2$, then clearly $|G_n| \leq n$. The extremal constructs are $K_{p,p}$ and the Hamiltonian cycle C_n.

(3.14). Since $t(K_{p,p+q}) = p < [(2p+q)/2]$, we have $t(G_n) \leq p$. Thus, $|G_n| \leq f(p+q,p)$. If $p+q \geq 2p+1$, or $q \geq p+1$, then $f(p+q,p) = p(p+q)$, and then (3.14) has been proved, since $K_{p,p+q}$ is a sufficient construct.

We note certain properties of a sufficient construct G_n. First we note that $\chi(G_n) = 2$. In fact, G_n does not have 'short' ($\leq 2p+q$) odd cycles which are not imbedded in $K_{p,p+q}$ as graphs. At the same time, the presence of a 'long' $(2p+q)$ cycle would imply the presence of more than p independent edges. Hence G_n can only contain the cycles $C_4, \ldots, C_{2p}$. If $\Delta(G_n) \leq p+q-1$, then $|G_n| \leq f(p+q-1,p) \leq p(p+q-1+1) = p(p+q)$, so that we assume that $\Delta(G_n) = p(p+q)$. Let l be the largest integer ($1 \leq l \leq p$) for which $K_{l,p+q} \subset G_n$, where the l-part of $K_{l,p+q}$ consists of the vertex set S_l and its $(p+q)$-part of S_{p+q}. Put $S_{n-l} = S_n \setminus S_l$, and consider the subgraph induced by these vertices: $G(S_{n-l}) \subset G_n$. Since $t(G_n) \leq p$, we have $t(G_{n-l}) \leq p - l$. If $\Delta(G(S_{n-l})) \leq p+q-1$, then

$$|G(S_{n-l})| \leq f(p+q-1, p-l) \leq (p-l)(p+q-1+l) = (p-l)(p+q),$$

and hence

$$|G_n| = l(p+q) + |G(S_{n-l})| \leq l(p+q) + (p-l)(p+q) = p(p+q).$$

Therefore we assume that $\Delta(G(S_{n-l})) = p+q$ and that $a \in S_{n-l}$ is a vertex having this degree in $G(S_{n-l})$. Since $\Delta(G(S_n)) = p+q$, we have $a \notin S_{p+q}$.

Consider the set $Z(a)$. Now $Z(a) = S_{p+q}$ contradicts the maximality of l, so that $Z(a) \cap (S_{n-l} \setminus S_{p+q}) \neq \emptyset$. Consider an edge $(a,b) \in G(S_{n-l})$ with $b \in Z(a) \cap (S_{n-l} \setminus S_{p+q})$. If now $1 \leq l \leq p-2$, then $K_{p,p+q} \not\supset K_{l,p+q} + (a,b)$, i.e. in this case also there cannot be a vertex of degree $p+q$ in $G(S_{n-l})$. We consider the remaining cases $l = p, p-1$. Since $|G(S_{n-l})| \leq f(p+q, p-l)$, we find (since $p+q \geq 2(p-l)+1$ for $l = p, p-1$) that in these cases the equality $f(p+q, p-l) = (p-l)(p+q)$ holds, and hence

$$|G_n| = |K_{l,p+q}| + |G(S_{n-l})| \leq l(p+q) + (p-l)(p+q) = p(p+q).$$

The extremal construct in case (3.14) is $K_{p,p+q}$.

(3.12). This case has been realized; the extremal construct is $K_{p,p+q}$.

In the cases (3.10)–(3.12) a general estimate holds: If $p \geq 2$, $q = 1$, then

$$\mu(n; K_{p,p+1}) \leq \max\left((p+1)^2, \left[\tfrac{3n}{2}\right]\right). \tag{3.15}$$

In fact, with the notation of the proof of (3.9), if $\Delta \geq 4$, then $t(G(S)) \leq p - \Delta + 1$, since $K_{p,p+1} \not\supset K_{1,\Delta} + F_{2(p-\Delta+2)}$ (which is true if $2p+1 \geq 1 + \Delta + 2p + 4 - 2\Delta$, or $\Delta \geq 4$), and hence $G(S) \not\supset F_{2(p-\Delta+2)}$. Consequently,

$$|G_n| \leq \Delta^2 + |G(S)| \leq \Delta^2 + f(\Delta, p-\Delta+1) = \Delta^2 + (p-\Delta+1)(\Delta+1) =$$
$$= p\Delta + p + 1 \leq (p+1)^2.$$

If $\Delta \leq 3$, then clearly $|G_n| \leq [3n/2]$. Thus, (3.15) has been proved.

For $n \geq 6$, (3.15) takes in the cases (3.10)–(3.11) the form

$$\mu(n; K_{p,p+1}) \leq \left[\tfrac{3n}{2}\right].$$

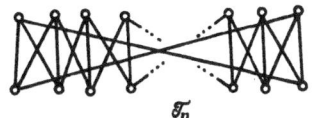

FIGURE 3.8. $\mathcal{T}_n$

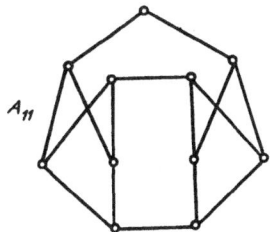

FIGURE 3.9. $\mathcal{A}_{11}$

(3.10). Let G_n be the extremal construct. The case $n = 5$ is trivial: $G_5 = K_{2,3}$. For $n = 7$, if $G_7 \supset C_7$, then $|G_7| = 7$; if $\chi(G_n) = 2$, then $|G_n| \leq 3[n/2]$, and this bound is realized by $K_{3,3}$. Let $n = 9$ and consider the shortest odd cycle contained in G_9. If it is C_9, then $|G_9| = 9$; if it is C_7, then for realizing (3.15), the remaining two vertices must be incident to six edges, which is impossible. Hence, here also $\chi(G_9) = 2$, i.e. $|G_9| \leq 3[9/2] = 12$, and this value is realized by the graph $\mathcal{T}_8$.

(3.11). If n is odd, then (3.15) is realized by $\mathcal{T}_n$ (see Figure 3). If $n = 11$, then (3.15) is realized by the graph $\mathcal{A}_{11}$. If $n \equiv 1 \pmod{4}$, $n \geq 13$, then (3.15) is realized by $\mathcal{A}_1$. If $n \equiv 3 \pmod{4}$, $n \geq 15$, then (3.15) is realized by $\mathcal{A}_3$. (See Figures 3– 3.) The sufficiency of these constructs can be verified immediately, and is based on the fact that for $p = 2$, $q = 1$ a graph G_n is a sufficient construct if and only if $\Delta(G_n) \leq 3$, $G_n \not\supset C_3, C_5$.

(3.13). Silvester's theorem, given in Chapter 2, substantiates the existence of the following graph, for nonnegative integers a and b,

$$\mathcal{A}_n = \begin{cases} K_{p+1,p+1} & \text{if } 3\left[\frac{n}{2}\right] \leq (p+1)^2, \\ a\mathcal{T}_6 + b\mathcal{T}_8 & \text{if } 3\left[\frac{n}{2}\right] \geq (p+1)^2,\ 3a + 4b = \left[\frac{n}{2}\right], \end{cases}$$

where $\mathcal{T}_6 = K_{3,3}$ and $\mathcal{T}_8$ is isomorphic to the edge graph of the ordinary three-cube.

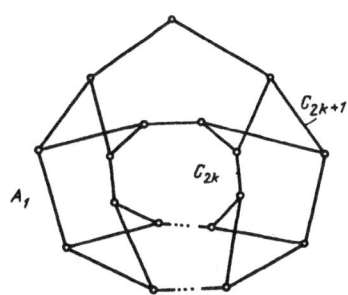

FIGURE 3.10. $\mathcal{A}_1$

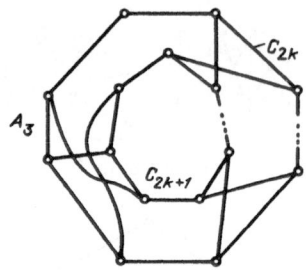

FIGURE 3.11. $\mathcal{A}_3$

We show that $\mathcal{A}_n$ is a maximal construct. Since $|\mathcal{A}_n| = \max((p+1)^2, 3[n/2])$, for n even it coincides with (3.15), so that (3.13) is proved for n even. Let n be odd. We show that $\mu(n; K_{p,p+1}) \leq \max((p+1)^2, 3[n/2])$. For this we consider an arbitrary extremal graph G_n with $\Delta(G_n) \leq 3$. It contains odd cycles, and exactly $[3n/2]$ edges (if one of these conditions is not satisfied, the required estimate follows at once). In turn, the condition $|G_n| = [3n/2]$ implies the existence in G_n of exactly $n-1$ vertices of degree 3 and one vertex of degree 2. Consider the odd cycle $C_r \subset G_n$. Clearly, $r > 2p + 1$. Consider a path $P_{2p+1} \subset C_r$ not passing through the degree-2 vertex of G_n (this can always be done, even if this vertex belongs to C_r, since $r > 2p + 1$).

For the sake of being specific, let

$$P_{2p+1} = \{(a_1, b_1), (b_1, a_2), (a_2, b_2), \ldots, (b_p, a_{p+1})\}$$

and $S_p = \{b_1, \ldots, b_p\}$ and $S_{p+1} = \{a_1, \ldots, a_{p+1}\}$. Consider all edges of G_n incident with vertices b_j and not belonging to the path P_{2p+1}. Since the cycle C_r is maximal, it does not contain chords, hence the number of such edges is p. We denote the set of ends of these edges by $V = \{v_1, v_2, \ldots\}$, and the graph formed by it and the path P_{2p+1} by H. If V contains a vertex v such that $(v, b_i), (v, b_j) \in H$ with $j - i > 1$, then the subgraph of H induced by the vertex set $S_{p+1} + (S_p \setminus b_l) + \{v\}$, where $i < l < j$, is connected and, being colorable by two colors, has $p - 1$ vertices of one color and $p + 2$ vertices of the other color, i.e. cannot be included in $K_{p,p+1}$. In the opposite case all vertices of H have degree at most 2, and if $(v, b_i), (v, b_j) \in H$, then $|i - j| = 1$, i.e. vertices of degree 2 are incident by 'neighboring' b_i.

Here we consider two cases. First, let $p \geq 4$ and let, for the sake of being specific, $(v_1, b_1), (v_2, b_{p-1}) \in H$. It is clear that then $v_1 \neq v_2$ (since the subgraph of H induced by the vertex set $(S_{p+1} \setminus \{a_{p+1}\}) + (S_p \setminus \{b_p\}) + \{v_1\} + \{v_2\}$ is a $(2p+1)$-vertex tree); it cannot be included in $K_{p,p+1}$, since it has $p - 1$ vertices of one color and $p + 2$ vertices of the other color. Finally, let $p = 3$. If $|V| = 3$, we can act as in the case $p \geq 4$. If $|V| = 2$, then the similar tree not imbeddable in $K_{3,4}$ is induced by either the vertex set $\{v_1, v_2, b_1, b_2, a_1, a_2, a_3\}$ or the vertex set $\{v_1, v_2, b_2, b_3, a_4, a_3, a_2\}$. $\square$

We[1] say that a graph G is an H-graph if any subgraph generated by $|V(H)|$ vertices contains H as a skeleton subgraph. Let $m(n, H)$ be the smallest possible number of edges in an n-vertex H-graph. In Theorem 3.2 we have shown that for a

---

[1]The following material (on local properties of hypergraphs) was written in collaboration with A.V. Kostochka.

graph H which is a k-cycle, for any $n \geq k$ we have

$$m(n, H) = \left\lceil \frac{n(n - k + 2)}{2} \right\rceil. \tag{3.16}$$

In this case the lower bound is trivial, since the minimal degree $\delta(G')$ of any k-vertex subgraph G' in a H-graph cannot be less than two. Of course, this very lower bound holds for $m(n, H)$ for any k-vertex H with minimal degree equal two. We show that (3.16) is true for $n > 8k^2$ for any 2-homogeneous k-vertex graph H, if there is at least one cycle of length larger than three in it. Moreover, we show that an n-vertex graph $G_{n,k}$ with $\lceil n(n - k + 2)/2 \rceil$ edges is an H-graph for every 2-homogeneous k-vertex graph, except if this k-vertex graph consists of 3-cycles only.

Below we set $N_G(v) = \{w \in V(G): (v, w) \in E(G)\}$, $\deg_G(v) = |N_G(v)|$.

Since for even n the construction is very simple, we regard this case separately.

LEMMA 3.3. *Let $G = (X, Y; E)$ be a bipartite graph without 4-cycles, $|X| \geq 2$, $|Y| \geq 3$, $|X \cup Y| \geq 7$. Then*

$$\exists x \in X : \deg_G(x) \leq |Y| - 2$$

and, moreover, either a)

$$\exists x_1, x_2 \in X, \, y_1, y_2 \in Y : \, x_1 \neq x_2, \, y_1 \neq y_2, \, (x_1, y_1), (x_2, y_2) \notin E(G), \tag{3.17}$$

or b) $|X| = 2$ *and some vertex in X is adjacent to all vertices in Y.*

PROOF. Let $X = \{x_1, \ldots, x_s\}$, $Y = \{y_1, \ldots, y_t\}$, $\deg_G(x_1) \leq \deg_G(x_2) \leq \cdots \leq \deg_G(x_s)$.

Case 1. $t \geq 4$. If $\deg_G(x_1) \geq |Y| - 1$, then $|N_G(x_1) \cap N_G(x_2)| \geq t - 2 \geq 2$, and G contains a 4-cycle passing through x_1 and x_2. Hence a) holds and there are $y_0, y_1 \in Y \setminus N_G(x_1)$.

Assume that b) does not hold. if there is a $y \in Y \setminus N_G(x_2)$, then we may assume that $y \neq y_1$ and, taking $y_2 = y$, we obtain a tuple x_1, x_2, y_1, y_2 satisfying (3.17). Hence $N_G(x_2) = Y$. Since G does not contain 4-cycles, we thus find $s = 2$.

Case 2. $t = 3$. Since $|X \cup Y| \geq 7$, we have $s \geq 4$. Since b) is symmetrical with respect to x and y, it has already been dealt with in Case 1.

Assume that a) is not true. Then

$$\deg_G(x_i) \geq 2, \qquad \forall 1 \leq i \leq s.$$

But for a set of three elements there are only three distinct pairs of vertices. Hence, because $s \geq 4$, we find $x_1 \neq x_2$ with $N_G(x_1) \cap N_G(x_2)| \geq 2$, i.e. G contains a 4-cycle passing through x_1 and x_2. □

THEOREM 3.4. *Let H be a 2-homogeneous graph on k ($k \geq 7$) vertices, in which the length of a maximal cycle is larger than 3. Then for $n \geq 8k^2$ the smallest number of edges in an n-vertex H-graph is equal to $\lceil n(n - k + 2)/2 \rceil$.*

PROOF. As noted above, it suffices to construct for each $n \geq 8k^2$ a graph $G = G_{n,k}$, with n vertices and $\lceil n(n - k + 2)/2 \rceil$ edges, in which any subgraph generated by k vertices contains all of the sets H considered as a skeleton subgraph.

Part 1. n even. For $n \geq 8k^2$ there is (see, e.g., [108]) a $(k - 3)$-homogeneous bipartite n-vertex graph $G_1 = (X, y; E_1)$ with $|X| = |Y|$ and without 4-cycles. Put

$G = \overline{G}_1$ (i.e. G is the completion of G_1). Clearly, G is an $(n - k + 2)$-homogeneous graph. Let $V' \subset X \cup Y$, $|V'| = k$, $X' = X \cap V'$, $Y' = Y \cap V'$. Since X and Y have equal cardinalities, we may assume that

$$|X'| \le |Y'|.$$

Then $|Y'| \ge 4$. Let H be an arbitrary 2-homogeneous graph from the conditions of the theorem. If $X' = \emptyset$, then $G' = G(Y')$ is a complete graph and $H \subset G'$.

Case 1. $X' = \{x\}$. Since $\deg_G(x) = n - k + 2$, there are $y_1, y_2 \in Y'$ adjacent to x. Since $G(Y')$ is a complete graph, this implies the inclusion $H \subset G'$.

Case 2. $X' = \{x_1, x_2\}$. Since x_1 and x_2 are adjacent to at least one vertex in Y', by Lemma 3.3 there are $y_1 \ne y_2$ in Y' such that $(x_1, y_1), (x_2, y_2) \in E(G')$. By requirement, H contains a cycle C_1 of length larger than 3. It can be included into G', using the path (y_1, x_1, x_2, y_2) and (if $|V(C_1)| > 4$) vertices in $Y' \setminus \{y_1, y_2\}$. The remaining cycles in H can clearly be included into the remaining complete subgraph of G'.

Case 3. $3 \le |X'| \le |Y'|$. Suppose H consists of the cycles $C_1, \ldots, C_r$, $|C_i| = c_i$ $(i = 1, \ldots, r)$. Let j be the index for which

$$c_1 + \cdots + c_{j-1} \le |X'| < c_1 + \cdots + c_j,$$

and let

$$b = |X'| - c_1 - \cdots - c_{j-1}, \qquad a = c_j - b.$$

If $b = 0$, then $C_1, \ldots, C_{j-1}$ can be included into X' and $C_j, \ldots, C_r$ into Y'.

If $b = 1$, then by Lemma 3.3 we can find an $x \in X'$ adjacent in G' to certain vertices y_1, y_2 from Y'. Distribute $C_1, \ldots, C_{j-1}$ onto $X' \setminus \{x\}$, C_j by using the path (y_1, x, y_2), and $C_{j+1}, \ldots, C_r$ onto the remaining vertices of Y'.

If $a = 1$, then we proceed similarly to the case $b = 1$.

If both $a, b > 1$, then we fix vertices $x_1, x_2 \in X'$, $y_1, y_2 \in Y'$ with $(x_1, y_1), (x_2, y_2) \in E(G')$ (these exist according to Lemma 3.3, and then place $C_1, \ldots, C_{j-1}$ into $X' \setminus \{x_1, x_2\}$, $C_{j+1}, \ldots, C_r$ into $Y' \setminus \{y_1, y_2\}$, and C_j onto the remaining vertices such that the edges (x_1, y_1) and (x_2, y_2) are used. This proves Part 1.

To prove Part 2 we need the following

LEMMA 3.4. *Let* $|V(G)| = p \ge 4$, $\delta(G) = p - 2$, $\{y_1, y_2, y_3, y_4\} \subset V(G)$, $(y_1, y_2), (y_3, y_4) \notin E(G)$. *Then for any* $c_0, \ldots, c_t$ *such that*

$$c_0 + \cdots + c_t = p \qquad and \qquad c_i \ge \begin{cases} 2, & i = 0, \\ 3, & 1 \le i \le t, \end{cases} \tag{3.18}$$

G *contains a subgraph consisting of a path* P_0 *and cycles* $C_1, \ldots, C_t$ *such that*

i) $|V(P_0)| = c_0$;
ii) $|V(C_i)| = c_i$ $(i = 1, \ldots, t)$;
iii) P_0 *joins* y_1 *with* y_3.

PROOF. For $p = 4$ the only choice satisfying (3.18) is $c_0 = 4$.

Let $p > 4$, and let $y_1, \ldots, y_4$ satisfy the conditions of the lemma. For $c_0 = 2$ we consider $G' = G \setminus \{y_1, y_3\}$. If $p - 2 = 3$, then, since y_2 and y_4 are isolated vertices in $\overline{G}'$, we have $G' = K_3$. If $p - 2 > 3$, $|E(\overline{G}')| \le 1$, the inclusion of $C_1, \ldots, C_t$ into G' is obvious. If, however, $|E(\overline{G}')| \ge 2$, then we take two arbitrary edges

$(y_5, y_6), (y_7, y_8) \in E(\overline{G}')$ and apply the induction hypothesis for $t' = t - 1$, $c'_i = c_{i+1}$ $(i = 0, \ldots, t-1)$, $y'_j = y_{j+4}$ $(j = 1, \ldots, 4)$.

Let $c_0 > 2$. If $E(\overline{G})| \geq 3$, then we choose in $\overline{G} \setminus \{y_1, \ldots, y_4\}$ an edge (y_5, y_6) and apply the assertion of the lemma to $G'' = G \setminus \{y_1\}$, with $c''_0 = c_0 - 1$, $c''_i = c_i$ $(i = 1, \ldots, t)$, $y''_j = y_{j+2}$ $(j = 1, \ldots, 4)$. If, however, $E(\overline{G}) = \{(y_1, y_2), (y-3, y_4)\}$, then the assertion of the lemma is obvious. $\square$

Part 2. n odd. For $n \geq 8k^2$ there is a bipartite $(k-2)$-homogeneous $(n+1)$-vertex graph $G_1 = (X_1, Y_1; E_1)$ without 4-cycles. Let v be a vertex from X_1, $N_{G_1}(v) = \{y_1, \ldots, y_{k-2}\}$, $G_2 = (G_1 \setminus \{v\}) \cup \{(y_{2i-1}, y_{2i}) : 1 \leq i \leq (k-2)/2\}$. Since the girth of G_1 is less than 6, the girth of G_2 is less than 5. Moreover,

$$|\tilde{Y} \cap N_{G_2}(x)| \leq 1, \qquad \forall x \in X_1 \setminus \{v\}, \tag{3.19}$$

where $\tilde{Y} = \{y_1, \ldots, y_{k-2}\}$. Put $G = \overline{G_2} = (V, E)$, where $V = X \cup Y$, $X = X_1 \setminus \{v\}$, $Y = Y_1$. By construction,

$$|E| = \binom{n}{2} - \left[\frac{(n+1)(k-2)}{2} - (k-2) + \left\lfloor \frac{k-2}{2} \right\rfloor \right] = \left\lceil \frac{n(n-k+2)}{2} \right\rceil.$$

As in Part 1, let $V' \subset V$, $|V'| = k$, $X' = X \cap V'$, $Y' = Y \cap V'$. If $G(Y')$ is a complete graph, we reason as in Part 1. Thus, we may assume that

$$\{y_1, y_2\} \subset Y', \qquad (y_1, y_2) \notin E, \qquad \deg_{G'}(y_1) \geq \deg_{G'}(y_2). \tag{3.20}$$

Condition (3.19) is equivalent to the following:

$$|\tilde{Y} \cap N_G(x)| \geq |\tilde{Y}| - 1, \qquad \forall x \in X. \tag{3.19'}$$

Case 1. $Y' = \{y_1, y_2\}$. Then $|X'| \geq 5$. Since $\delta(G') \geq 2$, we can find $x_1, x_2 \in X'$ adjacent to y_2. By (3.20), (3.19') there are $x_3, x_4 \in X' \setminus \{x_1, x_2\}$ adjacent to y_1. Since $G(X')$ is a complete graph, we can easily include H into $G(X' \cup \{y_1, y_2\})$, using the paths (x_1, y_2, x_2) and (x_3, y_1, x_4).

Case 2. $Y' = \{y_1, y_2, z\}$ Since $\delta(G') \geq 2$, we can find $x_1, x_2 \in X'$ such that $(x_1, y_1), (x_2, y_2) \in E$. By (3.19') we can choose $x_1 \neq x_2$. Thus, if H contains at least one cycle of length larger than 4, then H can be included into G', using the path (x_1, y_1, z, y_2, x_2). Consequently, H consists of cycles of lengths 4 and (possibly) 3. If there is an $x \in X'$ adjacent to y_1 and y_2, then the 4-cycle (x, y_1, z, y_2) and the complete graph $G(X' \setminus \{x\})$ show that it is possible to include H into G'. Thus, using (3.19') we have

$$X' \setminus N_{G'}(y_1) = X' \cap N_{G'}(y_2). \tag{3.21}$$

As noted, there is an $x_2 \in X'$ with $(x_2, y_2) \in E$. Assume that $|X' \cap N_{G'}(y-1)| \geq 3$. Because of (3.21) and the absence of a 4-cycle in $\overline{G}$, we have $|X' \cap N_{G'}(y_1) \cap N_{G'}(z)| \geq 2$. Let $x_1 \in X' \cap N_{G'}(y_1) \cap N_{G'}(z)$, $\{x_3, x_4\} \subset (X' \cap N_{G'}(y_1)) \setminus \{x_1\}$. Then the 4-cycle (x_2, y_2, z, x_1) and the path (x_3, y_1, x_4) show that it is possible to include H into G'. Thus, $|N_{G'}(y_1) \cap X'| \leq 2$. Since $k \geq 7$, by (3.20) this means that $k = 7$, $|X'| = 4$, $|X' \cap N_{G'}(y_1)| = |X' \cap N_{G'}(y_2)| = 2$, $H = C_3 \cup C_4$. Let $N_{G'}(y_1) \setminus \{z\} = \{x_1, x_3\}$, $X' \setminus \{x_1, x_3\} = \{x_2, x_4\}$ Since $\overline{G}$ does not contain 4-cycles, $\{x_1, x_3\} \cap N_{G'}(z) \neq \emptyset$. Let $(x_1, z) \in E$. Then G' contains the cycles (x_1, z, y_1) and (y_2, x_2, x_3, x_4).

Case 3. $|Y'| \geq 4$. Suppose H consists of the cycles $C_1, \ldots, C_r$, $|V(C_i)| = c_i$ $(i = 1, \ldots, r)$, $c_1 \leq \cdots \leq c_r$, and let j be such that

$$c_1 + \cdots + c_{j-1} \leq |Y'| < c_1 + \cdots + c_j.$$

Put $b = |Y'| - c_1 - \cdots - c_{j-1}$, $a = c_j - b$. If $b = 0$, then by Lemma 3.4 we can include $C_1, \ldots, C_{j-1}$ into $G(Y')$, and $C_j, \ldots, C_r$ can be trivially included into the complete graph $G(X')$.

Subcase 3.1. $E(\overline{G}(Y')) = \{(y_1, y_2)\}$, $z_1, z_2 \in Y' \setminus \{y_1, y_2\}$. First, suppose $a = 1$. If $|X'| > 1$, i.e. $j < r$, then $|X'| \geq 5$. Since $\overline{G}$ does not contain 4-cycles, we can then find an $x \in X'$ with $N_G(x) \cap \{z_1, z_2\} \neq \emptyset$. By (3.19'), $N_G(x) \cap \{y_1, y_2\} \neq \emptyset$. Assume that $\{(y_1, x)(x, z_1)\} \subset E$. Then it is easy to include $C_1, \ldots, C_j$ into $G(Y' \cup \{x\})$, using the path (y_1, x, z_1), and $C_{j+1}, \ldots, C_r$ can be included into the complete graph $G(X' \setminus \{x\})$. However, if $X' = \{x\}$, then by (3.20), $(x, y_1) \in E$ and since $\delta(G') \geq 2$ we can find a $y \in Y' \setminus \{y - 1\}$ with $(x, y) \in E$. Thus, we can include C_r into G', using the path (y, x, y_1) (recall that $c_r \geq 4$), and we can include $C_1, \ldots, C_{r-1}$ into the remaining complete subgraph $G' \setminus V(C_1)$.

Suppose now that $b = 1$. Then $|X'| \geq c_r - 1 \geq 3$ and, by (3.20), $|N_G(y_1) \cap X'| \geq 2$. Taking into account that $(x_1, y_1), (x_2, y_2) \in E$, we can place C_j into $X' \cup \{y_1\}$, using the path (x_1, y_1, x_2), and we can place $C_1, \ldots, C_{j-1}$ into the complete graph $G(Y' \setminus \{y_1\})$, and $C_{j+1}, \ldots, C_r$ into the remaining complete graph $G(X' \setminus V(C_j))$.

Finally, let $a, b \geq 2$, $\{x_1, x_2\} \subset X'$. Since $\overline{G}$ does not contain 4-cycles, g contains an edge joining $\{x_1, x_2\}$ and $\{z_1, z_2\}$. We may assume that $(x_1, z_1) \in E$ and, by (3.19'), $(x_2, y_1) \in E$. Then we can easily include $C_1, \ldots, C_{j-1}$ and a path of length $b - 1$ joining y_1 and z_1 into $G(Y')$, and $C_{j+1}, \ldots, C_r$ and a path of length $a - 1$ joining x_1 and x_2 into $G(X')$.

Subcase 3.2. $|E(\overline{G}(Y'))| \geq 2$. Let $\{y_3, y_4\} \subset Y' \setminus \{y_1, y_2\}$, $(y_3, y_4) \notin E$.

For $a = 1$ we choose an $x \in X'$ at random. Because of (3.19') we may assume that $\{y_1, y_3\} \subset N_G(x)$. Thus, by Lemma 3.4 we can include $C_1, \ldots, C_j$ into $G(Y' \cup \{x\})$, using the path (y_1, x, y_3).

Let $a, b \geq 2$, $\{x_1, x_2\} \subset X'$. Again because of (3.19') we may assume that $(x_1, y_1), (x_2, y_3) \in E$. By Lemma 3.4, we can include $C_1, \ldots, C_{j-1}$ and a path of length $b - 1$ joining y_1 and y_3 into $G(Y')$. We can include the cycles $C_{j+1}, \ldots, C_r$ and a path of length $a - 1$ joining x_1 and x_2 into the complete graph $G(X')$.

Only the variant $b = 1$ remains to be investigated. As in subcase 3.1, here we have $|X' \cap N_G(y_1)| \geq 2$, and by Lemma 3.4 the cycles $C_1, \ldots, C_{j-1}$ can be included into $G(Y' \setminus \{y_1\})$, unless $|Y'| = 4$, $j = 2$, $c_1 = 3$. The cycles $C_j, \ldots, C_r$ can be trivially included into $G(X' \cup \{y_1\})$. So, let $Y' = \{y_1, y_2, y_3, y_4\}$, $c_1 = 3$. Renumber the cycles such that $c_1 = \max\{c_i: 1 \leq i \leq r\}$. We obtain one of the variants already considered: $b = 0$, $a = 1$ or $a, b \geq 2$. This proves the theorem. $\square$

REMARK 3.1. It can be seen from the proof that $G_{n,k}$ is an H-graph not only for 2-homogeneous, but also for a whole series of other graphs H with $\delta(H) = 2$.

The local properties of graphs give rise to many very difficult extremal problems. Here we consider only one such property: the local Turán property. We say that an n-vertex k-graph $G_n^k \subset C^k(S_n)$ has the *local Turán property* if

$$\forall S_p \subset S_n \, \exists S_q \subset S_p : \; C^k(S_q) \subset G_n^k.$$

The main problem is to compute the quantities

$$m(n, p, q, k) = \min |G_n^k|,$$

where the minimum is over all locally Turán n-vertex k-graphs G_n^k. The relation between the Turán and local Turán properties is obvious: $T(n, p, q) = m(n, p, q, q)$. However, their interrelations are not exhausted by the fact that one is a particular case of the other. For example, the definitions imply that if a q-graph F^q has the Turán property, then the k-graph

$$G^k = \bigcup_{S_q \in F^q} C^k(S_q)$$

has the local Turán property.

THEOREM 3.5. *The following formulas hold:*

1) *if $p \le k(q-1)/(k-1)$, then $m(n, p, q, k) = \binom{n - \frac{p+q}{k}}{k}$;*

2) *if $n \le q(p-1)/(q-1)$, $2k \ge q+1$, then $m(n, p, q, k) = (n - p + 1)\binom{q}{k}$.*

PROOF. Let G^k have the local Turán property, and let the set

$$H \in \sum_{i=0}^{k(p-q)} C^i(S_n)$$

be such that

$$\forall S \in \sum_{i=0}^{k} C^i(S_n \setminus H) \, \exists S_p \supset H + S \, \exists S_q \subset S_p : \, C^k(S_p) \subset G^k, \, S \subset S_q.$$

To prove the existence of such a set H we consider a sequence of subsets $F = \{S^{(i)}\}_{1 \le i \le l}$ according to the following rule:

$$S^{(1)} = S_k \in C^k(S_n) \setminus G^k,$$

$$S^{(i)} \in \sum_{j=1}^{k} C^j \left(S_n \setminus \sum_{v=1}^{i-1} S^{(v)} \right) : \, \forall S_p \supset \sum_{j=1}^{i} S^{(j)} \, \forall S_p \subset S_q,$$

$$C^k(S_q) \subset G^k \Rightarrow S^{(i)} \not\subset S_q.$$

Suppose this process breaks off at level (step) l. We put

$$H = \sum_{i=1}^{l} S^{(i)}$$

and show that $l \le p-q$. Then, by construction, $|S^{(i)}| \le k$ and thus $|H| \le k(p-q)$, so that H clearly has the required properties. If $l = p-q+1$, then $|H| \le k(p-q+1) \le p$ by the sufficiency of G^k:

$$\forall S_p \supset H \, \exists S_q \subset S_p : \, C^k(S_q) \subset G^k.$$

Let S_p and S_q be precisely such. Then, by construction,

$$\forall S^{(i)} \in F : \, S^{(i)} \not\subset S_q,$$

thus, $|(S_p \setminus S_q) \cap S^{(i)}| \geq 1$, and hence (by the pairwise disjointness of sets in F:

$$\forall S^{(j)}, S^{(i)} \in F : S^{(i)} \cap S^{(j)} = \emptyset),$$

we obtain a contradiction:

$$p - q = |S_p \setminus S_q| \geq \sum_{i=1}^{p-q+1} |(S_p \setminus S_q) \cap S^{(i)}| \geq p - q + 1.$$

It is obvious that

$$\forall S_q \subset S_p \supset H : |S_q \cap H| \geq |H| - (p - q).$$

Hence, if

$$v(S_p, S_q, G) = |\{e \in G : S_p \cap e = S_q\}|,$$

then by the sufficiency of G^k and the definition of H we have

$$v(S_n \setminus H, S, G^k) \geq \binom{|S|}{|S|}\binom{|S_q \cap H|}{k - |S|} \geq \binom{|H| - p + q}{k - |S|}.$$

Then

$$|G^k| = \sum_{S \subset S_n \setminus H} v(S_n \setminus H, S, G^k) \geq \sum_{S \subset S_n \setminus H} \binom{|H| - p + q}{k - |S|} =$$

$$= \sum_{j=0}^{k} \binom{n - |H|}{j}\binom{|H| - p + q}{k - j} = \binom{n - p + q}{k},$$

or $|G^k| \geq \binom{n-p+q}{k}$. Since the construct $C^k(S_{n-p+q})$ is sufficient, this implies the first assertion of the theorem.

We represent the minimal locally Turan k-graph G^k in the form

$$G^k = \bigcup_{S_q \in F^q} C^k(S_q),$$

where the Turan q-graph F^q has the following two properties: it is critical, i.e. by removing a q-edge of it, it no longer has the Turan property; it has the *property of pairwise separatedness*:

$$\forall S_q^{(i)}, S_q^{(j)} \in F^q : S_q^{(i)} \cap S_q^{(j)} = \emptyset.$$

Then

$$|G^k| \geq T(n, p, q)\binom{q}{p} = (n - p + 1)\binom{q}{k},$$

and if always

$$\exists S_q^{(i)}, S_q^{(j)} \in F^q : S_q^{(i)} \cap S_q^{(j)} \neq \emptyset,$$

then

$$\exists a \in S_n : v(a, 1; G^k) \geq \binom{q}{k}.$$

We show the presence of such a vertex. It is immediately clear that if

$$k \geq \frac{1}{2}((5q^2 - 8q + 4)^{1/2} - q + 2),$$

then there is such a vertex, since under these k conditions the presence of two intersecting q-edges $S_q^{(i)}, S_q^{(j)} \in F^q$ implies

$$\forall a \in S_q^{(i)} \cap S_q^{(j)} : v(a, 1; G^k) \geq 2\binom{q-1}{k-1} - \binom{|S_q^{(i)} \cap S_q^{(j)}| - 1}{k-1} \geq$$

$$\geq 2\binom{q-1}{k-1} - \binom{q-2}{k-1} > \binom{q}{k}.$$

Let now F^q be a critical Turan k-graph, and let

$$\forall S_q^{(i)}, S_q^{(j)}, S_q^{(l)} \in F^q : S_q^{(i)} \cap S_q^{(i)} \cap S_q^{(l)} = \emptyset.$$

If also

$$\exists S_q^{(i)}, S_q^{(j)} \in F^q : |S_q^{(i)} \cap S_q^{(j)}| \leq \tfrac{q}{2},$$

then for such q-edges,

$$\forall a \in S_q^{(i)} \cap S_q^{(j)} : v(a, 1; G^k) \geq \binom{q}{k}.$$

If, however,

$$\forall S_q^{(i)}, S_q^{(j)} \in F^q : \text{either } |S_q^{(i)} \cap S_q^{(j)}| > \frac{q}{2} \text{ or } S_q^{(i)} \cap S_q^{(j)} = \emptyset,$$

then F^q is clearly not critical. The alternative

$$\forall S_q^{(i)}, S_q^{(j)}, S_q^{(l)} \in F^q : S_q^{(i)} \cap S_q^{(i)} \cap S_q^{(l)} \neq \emptyset$$

implies the required immediately, provided

$$\forall S_q^{(i)}, S_q^{(j)}, \in F^q : \text{either } |S_q^{(i)} \cap S_q^{(j)}| = q - 1 \text{ or } S_q^{(i)} \cap S_q^{(j)} = \emptyset,$$

but the latter implies, by the Turan property and criticality of F^q, that

$$\exists S_{q+1} \subset S_n : C^q(S_{q+1}) \subset F^q.$$

Thus, for this S_{q+1},

$$\forall a \in S_{q+1} : v(a, 1, G^k) \geq \binom{q}{k}.$$

In turn, the presence of vertices of arbitrary high degree is guaranteed by induction with respect to the number of vertices. Clearly, $m(p, p, q, k) = \binom{q}{k}$ for any $p \geq q \geq k$. Suppose this equality holds up to $n - 1$ inclusive. We show that it holds also for n. Assume the contrary: $m(n, p, q, k) < (n - p + 1)\binom{q}{k}$. It is obvious that if the k-graph G^k is locally Turan, then the k-graph obtained from it by removing all k-edges incident to an arbitrary vertex is an $(n - 1)$-vertex k-graph, and is also locally Turan, with the same parameters. Suppose $a \in S_n$ is a vertex such that

$$v(a, 1; G^k) \geq \binom{q}{k}.$$

Then, by the preceding Remark and the induction hypothesis,

$$m(n-1,p,q,k) \le |G^k| - v(a,1;G^k) \le m(n,p,q,k) - \binom{q}{k} <$$

$$< (n-p+1)\binom{q}{k} - \binom{q}{k} = (n-p)\binom{q}{k},$$

contradicting the induction hypothesis. $\square$

4. Asymptotics for local properties of graphs

If it is known that a k-vertex graph G_k is contained as a subgraph in the complete t-partite graph with parts of k_i vertices ($k_1 \ge \cdots \ge k_t$, $k_1 + \cdots + k_t = k$), then, clearly, the following 'chromatic' information can be deduced:

$$\chi(G_k) \le t, \qquad \chi(\bar{G}_k) \ge k_1.$$

Can we extract 'chromatic' information concerning G_k if such an inclusion as above is not present? It turns out we can.

The *outer chromatic number* of a k-vertex graph F_k is the number

$$\chi'(F_k) = \min_{G_k \not\subseteq F_k} \chi(G_k),$$

where $\chi(G_k)$ is the ordinary chromatic number of the k-vertex graph G_k. Unlike the ordinary chromatic number, the outer chromatic number can be explicitly calculated.

LEMMA 3.5. *If $F_k \ne K_k$ and $\bar{F}_k$ has t connectedness components, with k_i vertices in component i and $k_1 \ge \cdots \ge k_t \ge 1$, $k_1 + \cdots + k_t = k$, then*

$$\chi'(F_k) = \min_{i:\, k_i > 1} \left\lceil \frac{\sum_{j=i}^{t} k_j - 1}{k_i - 1} \right\rceil + 1.$$

PROOF. If $G_k \not\subseteq F_k$ and $G_k \subseteq K_{n_1 \ldots n_r}$, where $(n_1, \ldots, n_r) \vdash k$, then $K_{n_1 \ldots n_r} \not\subseteq F_k$. Hence the equality $\chi'(F_k) = r$ is equivalent to

$$\exists(n_1, \ldots, n_r) \vdash k : K_{n_1 \ldots n_r} \not\subseteq F_k,$$
$$\forall(n_1, \ldots, n_{r-1}) \vdash k : K_{n_1 \ldots n_{r-1}} \subseteq F_k.$$

Thus, if r is the largest integer for which

$$\forall(n_1, \ldots, n_r) \vdash k : K_{n_1 \ldots n_r} \subseteq F_k,$$

then $\chi'(F_k) = r + 1$. In turn,

$$K_{n_1 \ldots n_r} \subseteq F_k \Leftrightarrow \bar{K}_{n_1 \ldots n_r} \supseteq \bar{F}_k,$$

or

$$K_{n_1} + \cdots + K_{n_r} \supseteq H = \bar{F}_k.$$

The latter inclusion clearly holds if and only if $(n_1, \ldots, n_r) \supseteq (k_1, \ldots, k_t) \vdash k$, where k_i is the number of vertices in the ith connectedness component of H_k. Consequently, the number r looked for is the number $r(k_1, \ldots, k_t; k)$ from Corollary 2.1, according to which (taking into account that $F_k \neq K_k$ and, thus, $k_1 > 1$):

$$r(k_1, \ldots, k_t; k) = \min_{i: k_i > 1} \left[\frac{k - \sum_{j=1}^{i} k_j}{k_i - 1} \right] + 1 =$$

$$= \min_{i: k_i > 1} \left[\frac{\sum_{j=i+1}^{t} k_j}{k_i - 1} \right] + 1 = \min_{i: k_i > 1} \left[\frac{\sum_{j=i}^{t} k_j - 1}{k_i - 1} \right].$$

This proves the lemma. $\square$

We now note a relation between the ordinary and outer chromatic numbers. The inequality

$$\chi(G_k) \geq \max_{F_k \not\supseteq G_k} \chi'(F_k)$$

follows immediately from the definition of outer chromatic number. In particular, it allows us to obtain an explicit lower bound for the ordinary chromatic number. In particular, we have

$$\chi(G_k) \geq \max \min_{i: k_i > 1} \left[\frac{\sum_{j=i}^{t} k_j - 1}{k_i - 1} \right] + 1,$$

where the maximum is taken over the partitions $(k_1, \ldots, k_t) \vdash k$ for which the graph $K_{k_1 \ldots k_t}$ does not contain the graph G_k as a subgraph. Therefore, the optimal bound is obtained for partitions $(k_1, \ldots, k_t) \vdash k$ maximizing the function $r(k_1, \ldots, k_t; k)$. This bound is well-known: If b is the largest number of independent vertices in the graph G_k, then $\chi(G_k) \geq k/b$. We derive it from the previous inequality. For this we put

$$H_k = K_{b+1} + (k - b - 1)K_1.$$

This means that the graph H_k consists of the complete graph K_{b+1} and $k - b - 1$ isolated vertices. Then

$$\bar{H}_k = F_k \not\supseteq G_k \quad \text{and} \quad \chi(G_k) \geq \chi'(F_k) = \left[\frac{k-1}{b} \right] + 1 \geq \frac{k}{b}.$$

Somewhat different concrete general bounds of the chromatic number in terms of the outer chromatic number are given in the Problem Section. Knowledge of the precise value of the outer chromatic number makes it possible to asymptotically solve problems on local properties, and to explicitly compute the coefficient in front of the principal asymptotic term.

THEOREM 3.6. *Suppose a nonempty k-vertex graph H_k consists of t connectedness components, with k_i vertices in component i and $k_1 \geq \cdots \geq k_t \geq 1$, $k_1 + \cdots + k_t = k$. Then*

$$m(n; H_k) = \frac{n^2}{2 \min_{i: k_i > 1} \left[\frac{\sum_{j=i}^{i} k_j - 1}{k_i - 1} \right]} + o(n^2).$$

PROOF. This theorem follows immediately from the Erdős–Shimonovich theorem and Lemma 3.5:

$$m(n; H_k) = \binom{n}{2} - \mu(n; F_k) = \binom{n}{2} - f(n; \{G_k : G_k \not\subset F_k\}) =$$

$$= \binom{n}{2} - \left(1 + \frac{1}{1 - \chi'(F_k)}\right)\frac{n^2}{2} + o(n^2) = \frac{n^2}{(\chi'(F_k) - 1)/2} + o(n^2) =$$

$$= \frac{n^2}{2\min_{i:\, k_i>1}\left[\frac{\sum_{j=1}^{i} k_j - 1}{k_i - 1}\right]} + o(n^2). \qquad \square$$

5. Elements of Ramsey theory

Already in Chapter 2 we have noted an extremal result established by Ramsey. Results of a similar nature are nowadays formulated in a separate branch, which quite often is called *Ramsey theory*. The problem theme in it can, in essence, be reduced to the two following problems.

If a large structure is partitioned into disjoint parts, then how many substructures are bound to belong to a single part? Conversely: How rich can the large structure be if any partition of it contains a part of a prescribed nature? The basic initial result of this type, which is also the basic instrument for solving similar problems, is the Dirichlet principle.

We illustrate this by the standard examples of Ramsean kind.

- Among any three persons there are two of the same sex.
- Among any six people there is either a triple of pairwise acquaintances or a triple of pairwise strangers.

While the first example is a simple reformulation of the box principle, the second is a particular case of Ramsey's theorem. Consider a proof of this example. For any person in this sextet of persons there are p acquaintances and q strangers among the remaining five, so that p and q take the values $0, 1, \ldots, 5$, with the condition $p + q = 5$. It is obvious that $\max(p, q) \geq 3$; for the sake of being specific we assume that $p \geq q$. Consider these p acquaintances. If there is at least one pair of mutual acquaintances among them, we have found a triple of acquainted persons; if there is no pair of acquaintances among these p persons, then we have a triple of pairwise strangers, since $p \geq 3$. This proves the required assertion.

We must note that the amount of six persons is extremal; in fact, it is the smallest number with the above property, since there is a selection of five persons not having this property, i.e. in which there is no triple of pairwise acquainted persons nor a triple of pairwise strangers. This results from the intransitivity of 'being acquainted', regarded as a binary relation, and the obvious fact that the complete five-vertex graph K_5 can be represented as the union of two disjoint pentagons: $K_5 = C_5 + C_5$, where the first cycle is the graph of pairwise acquaintances and the second that of pairwise strangers. This suggests that the relation 'being acquainted–being strangers' can be adequately represented by coloring the edges of the complete graph with two distinct colors, e.g. pairs of acquaintances with red and pairs of strangers by blue. Ramsey's theorem can be conveniently stated precisely in terms of edge colorings of graphs.

THEOREM 3.7 (RAMSEY THEOREM (PARTICULAR CASE)). *For natural numbers r and s there is a smallest integer $R = R(r,s)$ such that in any full edge coloring of the complete R-vertex graph K_R with two colors (with each edge colored by one of these two colors) there is either a complete subgraph K_r of the first color or a complete subgraph K_s of the second color.*

Thus, our example establishes that $R(3,3) = 6$. In general, the exact computation of such *Ramsey numbers* for graphs is a difficult and by far unsolved problem. The values of $R(r,s)$ for even the most natural parameter values are far from being known completely, as has been indicated in Table 3.1, which gives all presently known information concerning the exact values of the Ramsey numbers for graphs when coloring with two colors.

s \\ r	3	4	5	6	7	8	9	10	11	12	13	14	15
3	6	9	14	18	23	28	36	40 43	46 51	51 60	59 69	66 78	73 89
4		18	25 26	35 41	49 62	53 85	69 116	80 151	93 191	97 238	112 291	119 349	121 417
5			43 50	58 89	76 145	95 219	320						
6				102 166	304	499	786						
7					205 582	1048	1783						
8						282 1998	3675						
9							565 7134						
10								798					

Table 3.1. Estimates and exact values of the Ramsey numbers $R(r,s)$

We have taken this table from [206].[2]

Ramsey numbers may be computed not only for ordinary graphs, but also for l-graphs, and, moreover, more than two colors may be used.

By analogy with this result we may consider the packability of number partitions in terms of colorings of 1-graphs or 1-subsets, and it is therefore natural to extend this packability to l-graphs and to compare the corresponding extremal bounds with Ramsey numbers. Let $C(r,s)$ be the smallest integer C such that in any edge coloring of the complete C-vertex graph K_C with two colors there are edge-disjoint subgraphs K_r and K_s that are each of a single color. The following fact confirms that the difference between C and R for 2-subsets already is less striking than for 1-subsets, but that, on the other hand, these numbers coincide in essence.

PROPOSITION 3.1. *If $r > s$, then $C(r,s) = R(r,r)$.*

---

[2]On March 19, 1993 we received the following by e-mail from B.D. McKay and S.P. Radziszowski: 'We have proved that $R(4,5) = 25$. Our proof is computational. For integers s,t define an (s,t,n)-graph to be an n-vertex graph with no clique of order s or independent set of order t. Suppose that G is a $(4,5,25)$-graph, with 25 vertices. If a vertex is removed from G, a $(4,5,24)$-graph H results; moreover, the structure of H can be somewhat restricted by choosing which vertex of G to remove. Our proof consists of constructing all such structure-restricted $(4,5,24)$-graphs and showing that none of them extends to a $(4,5,25)$-graph. To reduce the chance of computational error, the entire computation was done in duplicate using independent programs written by each author. The fastest of the two computations required about 3.2 years of CPU time on Sun workstations.'

PROOF. Clearly, $C(r,s) > R(r,r)$ always, and if $R(r,r) - r + 1 \geq R(s,s)$, then $C(r,s) = R(r,r)$. In fact, put $R = R(r,r)$ and consider an arbitrary coloring of the complete R-vertex graph K_R with two colors. This coloring must include a one-colored graph K_r, say on the vertices $[r]$. Consider now the coloring on the remaining vertices $[r, R] = \{r + 1, \ldots, R\}$. Since $\|[r, R]\| = R - r + 1 \geq R(s,s)$, this coloring contains a complete s-vertex graph K_s, which can intersect the graph K_r in at most one vertex. Hence we have obtained the required configuration.

We now show that if $r > s$, then $R(r,r) - r + 1 \geq R(s,s)$ always. For this it suffices to give a coloring of $K_{R(s,s)+r-2}$ with two colors, not containing a one-colored complete graph on r vertices. Represent $K_{R(s,s)+r-2}$ as a sum of disjoint subgraphs:

$$K_{R(s,s)+r-2} = K_{R(s,s)-1} + K_{r-1} + K_{R(s,s)-1,r-1},$$

where the edges of $K_{R(s,s)-1}$ are colored with two colors such that this graph does not contain a one-colored complete subgraph on s vertices, all edges of K_{r-1} are colored blue, and all edges of $K_{R(s,s)-1,r-1}$ are colored red. It is obvious that with this coloring the graph $K_{R(s,s)+r-2}$ does not contain a one-colored complete subgraph on r vertices. $\square$

Not only hypergraphs can be taken as structures to be partitioned, but also totally different sets, e.g. of numerical or geometrical objects. The earliest number-theoretical facts in Ramsey theory are the following theorems.

THEOREM 3.8 (SCHUR'S THEOREM). *In any partition of the set of natural numbers in finitely many parts, there is a part containing numbers x, y, z such that $x + y = z$.*

THEOREM 3.9 (VAN DER WAERDEN THEOREM). *In any partition of the set of natural numbers into two parts, there is a part containing an arithmetical progression of l terms, $a, a + b, \ldots, a + (l - 1)b$, whatever the given finite length l.*

The extremal form of van der Waerden's theorem is the problem of computing $\mathcal{W}(n)$, the smallest integer $\mathcal{W}$ such that in any partition of the set of the first $\mathcal{W}$ natural numbers, $[\mathcal{W}] = \{1, \ldots, \mathcal{W}\}$, into two parts, there is a part containing an n-term arithmetical progression. This is an open problem, as is the computation of Ramsey numbers for graphs.

The best upper bound for $\mathcal{W}(n)$ is due to S. Selah [205].

Geometrical facts of Ramsean type emerged somewhat later. The earliest of them has been considered by Erdős and Szekeres, in the form of the following extremal problem posed by E. Klein.

Compute $N(n)$, i.e. the smallest integer N such that we can choose from N points in general position in the plane (i.e. no three lying on the same line) n points forming the vertices of a convex n-gon.

In particular, it has been proved that

$$2^{n-2} + 1 \leq N(n) \leq \binom{2n - 4}{n - 2} + 1,$$

and it has been conjectured that the lower bound is the exact value. This conjecture has been confirmed only for $n = 3, 4, 5$. It was precisely the estimation of $N(n)$ that led Erdős to the re-discovery of Ramsey's theorem, somewhat later than Ramsey himself, however.

In specific forms of Ramsey statements one considers, next to incidence relations, also special relations: in number-theoretical ones additive relations, and in geometrical ones configuration and metrical relations. The following is an example of calculation with metrical relations.

For any coloring of the points of the plane by three colors there is a pair of points of the same color having distance one. The problem on the smallest number of colors for which this property does not hold is open; it is only known that this number lies in between 4 and 7.

In the course of solving the problem of Klein on convex polygons, an extremal problem concerning permutations emerged (see Problem 3.25).

Nowadays, combinatorial permutation problems form a separate direction, consisting both of purely combinatorial 'permutational' statements, and of problems connected with group properties of permutations. This theme deserves special consideration, which would lead outside the scope of this book. We restrict ourselves to given some statements in the form of problems.

6. Problems and assertions

PROBLEM 3.1. Suppose we are given a hypergraph $F \subseteq \mathcal{P}(S_n)$ and a system of hypergraphs $\mathcal{W} = \{G, \dots\}$, $G \subseteq \mathcal{P}(S_n)$, on the vertex set S_n. Moreover, suppose we are given a binary relation $R \subseteq \mathcal{P}^2(S_n)$ on the Boolean $\mathcal{P}(S_n)$. For $G \subseteq \mathcal{P}(S_n)$ we denote by $R(G)$ the complete image of G under R, i.e. $R(G) = \{X \in \mathcal{P}(S_n): \exists e \in G: eRX\}$, and for $S \subseteq S_n$ we introduce the quantity

$$\deg_{\mathcal{W}}(S) = |\{G \in \mathcal{W} : S \in R(G)\}|.$$

Then

$$\sum_{S \in F} \deg_{\mathcal{W}}(S) = \sum_{G \in \mathcal{W}} |R(G) \cap F|.$$

PROBLEM 3.2. If G is a hypergraph and S a vertex set, then

$$\sum_{S_p \subseteq S} \binom{v(S_p, p; G)}{k} = \sum_{1 \le i_2 < \cdots < i_k \le |G|} \binom{|S \cap \bigcap_{j=1}^{k} e_{i_j}|}{p},$$

$$\sum_{S_p \subseteq S} v(S_p, q; G) = \sum_{e \in G} \binom{|e \cap S|}{q} \binom{|S| - |e \cap S|}{p - q}.$$

PROBLEM 3.3. Compute the sum

$$\sum_{S_p \subseteq S} \binom{v(S_p, q; G)}{k}.$$

PROBLEM 3.4. If a graph H_k consists of t connectedness components with l vertices in each component and $k - lt$ isolated vertices, then

$$m(n; H_k) = \left(\frac{1}{1 + \left\lceil \frac{k-lt}{l-1} \right\rceil} \right) \frac{n^2}{2} + o(n^2).$$

PROBLEM 3.5. If the graph H_k does not have isolated vertices, then

$$m(n; H_k) = \frac{n^2}{2} + O(n).$$

PROBLEM 3.6. *Forbidden subgraphs.*

$f(n; C_n) = \binom{n-1}{2} + 1$; the extremal construct is the complete $(n-1)$-vertex graph K_{n-1}, complemented by an edge incident to the nth vertex;

$$f(n; \{C_3, C_4, \ldots\}) = n - 1;$$

$$f(n; C_l) = \binom{l-1}{2} + \binom{n-l+2}{2}, \qquad l \le n \le 2l - 3;$$

$$f(n; C_{2l+1}) = \begin{cases} \binom{n}{2}, & n \le 2l, \\ \binom{2l}{2} + \binom{n-2l+1}{2}, & 2l \le n \le 4l - 1, \\ \left[\frac{n^2}{4}\right], & n \ge 4l - 1; \end{cases}$$

$$f(n; \{C_l, C_{l+1}, \ldots, C_n\}) = \frac{n(l-1)}{2} - \frac{r(l-r)}{2}, \qquad n = q(l-2) + r, \quad 0 < r \le l - 2;$$

$$f(n; \{C_4, C_6, C_8, \ldots\}) = n - 1 + \left[\frac{n-1}{2}\right].$$

In the last case the extremal construct is the graph $C^1(a)C^1(S_n \setminus \{a\}) + \mathcal{F}(S_n \setminus \{a\})$.

PROBLEM 3.7. *Forbidden subgraphs: the 'degenerate' case.*

$$f(n; C_r) \le c \cdot n^{1+1/r};$$
$$f(n; C_4) = 0.5 \, n^{1.5} + O(n);$$
$$f(n; \{C_4, C_5\}) = (0.5 \, n)^{1.5} + O(n);$$
$$c_1 \cdot n^{5/3} \le f(n; K_{3,3}) \le c_2 \cdot n^{5/3}.$$

PROBLEM 3.8. (?) The *Erdös—Sös conjecture.* For every k-vertex tree T_k we have $f(n; T_k) \le n(k-2)/2$.

PROBLEM 3.9. If $n \ge k \ge 2$ and k is even, then every n-vertex graph without isolated vertices contains a proper k-vertex subgraph without isolated vertices.

PROBLEM 3.10. Prove the following equation:

$$m(n; P_k) = \begin{cases} k - 1 & \text{if } n = k, \\ \left] \frac{n(n-k+1)}{2} \right[& \text{if } n > k. \end{cases}$$

PROBLEM 3.11. Let $M(n, k)$ be the minimum number of edges in a graph $G_n \subset C^2(S_n)$ such that

$$\forall S_k \subset S_n \exists a \in S_n \setminus S_k : \ C^1(a) \cdot C^1(S_k) \subset G_n.$$

Then

$$m(n, k) = (k-1)n - \binom{k}{2} + \left[\frac{n-1}{2}\right] + 1,$$

and the extremal construct has the form

$$C^2(S_n) - C^2(S_{n-k+1}) + \mathcal{F}'(S_{n-k+1}).$$

PROBLEM 3.12. Let $E(n,t)$ be the maximum number of edges in an n-vertex graph with number of independent edges at most t. Then

$$E(n,t) = \begin{cases} \binom{n}{2} & \text{if } n \le 2t+1, \\ \max\left(\binom{2t+1}{2}, \binom{n}{2} - \binom{n-t}{2}\right) & \text{if } n \ge 2t+1. \end{cases}$$

PROBLEM 3.13. If $\alpha_l(G_k)$ is the largest number of disjoint independent vertex l-sets in a graph G_k, then for any l such that $2 \le l \le b+1$ we have

$$\chi(G_k) \ge \left\lceil \frac{k - l\alpha_l(G_k) - 1}{l-1} \right\rceil + 1.$$

PROBLEM 3.14. If $\delta(G_k)$ is the smallest degree in a graph G_k, then

$$\chi(G_k) \ge \left\lceil \frac{k-1}{k - \delta(G_k)} \right\rceil + 1.$$

PROBLEM 3.15. The maximum number of k-edges in an n-vertex k-graph $G_n^k \subset C^k(S_n)$ not containing a triple of k-edges $A, B, C \in G_n^k$ such that $A \triangle B \subset C$ is, for $k = 2, 3, 4$, equal to

$$\left\lceil \frac{n}{k} \right\rceil \left\lceil \frac{n+1}{k} \right\rceil \cdots \left\lceil \frac{n+k-1}{k} \right\rceil.$$

PROBLEM 3.16. The minimal total number of triangles in an n-vertex graph and its complement is

$$\binom{n}{3} - \left\lceil \frac{n}{2} \left[\left(\frac{n-1}{2} \right)^2 \right] \right\rceil.$$

Hint. Show that the total number of triangles in and n-vertex graph and its complement can be expressed by

$$\frac{1}{2} \left(\sum_{i=1}^n \binom{d_i}{2} + \sum_{i=1}^n \binom{n-1-d_i}{2} - \binom{n}{3} \right),$$

where $d_1, \ldots, d_n$ are the vertex degrees of the graph. Thus, one has to minimize this expression, taking into account that the sum of all degrees in the graph is even.

PROBLEM 3.17. The *edge chromatic number* of a graph G is defined as the smallest integer X for which there is an edge coloring of this graph with X colors such that any two adjacent edges have different colors. Show that if Δ is the maximal degree of G and t is the largest number of independent edges in G, then $\Delta \le X \le \Delta+1$, while if $|G| > \Delta t$, then $X = \Delta + 1$, and if $\Delta \ge 2t + 1$, then $X = \Delta$.

PROBLEM 3.18. How many edges must an n-vertex graph have if between any k vertices there are t independent edges?

PROBLEM 3.19. (?) Try to compute $\mu(n, K_p + K_q)$.

PROBLEM 3.20. **a)** We call a graph countably-covering-connected if for any pair of vertices in it there is a system of countably many vertex-disjoint paths joining these vertices and covering all vertices of the graph. Show that a plane graph is Hamiltonian if and only if it is countably-covering-connected.

Hint. Use a sufficient condition for being Hamiltonian, due to Tait: a plane four-connected graph is Hamiltonian.

b) (?) Is the previous criterion for being Hamiltonian true for nonplanar graphs?

PROBLEM 3.21. Let n_a be the smallest n such that for any edge coloring of the complete graph K_n with two colors there are two one-color triangles, possibly not of the same color, but without common edges. Then $n_a = 7$.

PROBLEM 3.22. Let n_b be the smallest n such that for any edge coloring of the complete graph K_n with two colors there are two one-color triangles of the same color and without common edges. Then $n_b = 8$.

PROBLEM 3.23. Show that for the Ramsey numbers the following inequality holds:

$$R((r-1)(s-1)+1, (r-1)(s-1)+1) > (R(r,s)-1)(R(s,s)-1).$$

Hint. Consider $K_{R(r,r)-1}$, color it with two colors such that it does not contain a one-color subgraph on r vertices in which each 'vertex' is a complete graph $K_{R(s,s)-1}$, colored with two colors such that it does not contain a one-color subgraph on s vertices.

PROBLEM 3.24. Let $m(n,2,3)$ be the maximum number of hyperedges in n n-vertex hypergraph such that each two hyperedges have a nonempty intersection, while each tree hyperedges have empty intersection. Then

$$m(n,2,3) = \left[1 + (8n+1)^{0.5}2\right].$$

analyze the relation between m and the Turan numbers $T(n,k,l)$.

PROBLEM 3.25. How many terms can a maximal monotone subsequence in a rearrangement of the first n natural number have?

PROBLEM 3.26. What is the average (in the set of all rearrangements) length of a maximal monotone subsequence in a rearrangement of the first n natural numbers?

PROBLEM 3.27. *Universal key for naming files.* Assume that the unknown name of a file consists of n different symbols. How short can a subsequence of n symbols, in which all possible rearrangements of these n symbols are present as subsequences from subseries of distributed elements of the sequence be?

PROBLEM 3.28. *Universal ruler.* How many lines, $N(n)$, should be arranged on a ruler such that precisely any integer from 0 to n can be measured by these lines?

PROBLEM 3.29. *Local Ramsey property.* Let H_k be an arbitrary k-vertex graph with at least one edge, and let $m \geq 2$ be a natural number. Show that there is a smallest natural number $R = LR(H_k, m)$ $(R \geq k)$ such that for any R-vertex graph G_R one of the two following conditions holds:

$$\exists G_k \subset G_R : H_k \not\subset G_k \quad \text{or} \quad \mathrm{cl}(G_R) \geq m,$$

where $\mathrm{cl}(G)$ is the clique number of the graph G, i.e. the number of vertices of the largest complete subgraph in G.

Prove the following equations

(1) $LR(K_2 \cup (k-2)K_1, m) = R(k, m)$;

(2) $LR(P_3 \cup (k-3)K_1, m) = \max(k, R(K_k - \mathcal{F}_k, m))$;

(3) suppose $K_{1,k-1} \subseteq H_k$; then

$$LR(H_k, m) = \begin{cases} k - \mathrm{cl}(H_k) + \max\{m, \mathrm{cl}(H_k)\}, & \mathcal{F}_k \not\subset (K_k - H_k), \\ \max\{k, m + \max\{m-1, k - \mathrm{cl}(H_k)\}\}, & \mathcal{F}_k \subset (K_k - H_k) \end{cases}$$

(use Theorem 3.1);

(4) $LR(K_{2,3}, m) = 3m - 2 \ (m \geq 3)$

(concerning the solution of this problem see [59]).

PROBLEM 3.30. Let $R(G, H)$ be the smallest R for which any edge 2-coloring of the complete graph K_R has either a subgraph G of the first color or a subgraph H of the second color. (The existence of $R(G, H)$ follows from Ramsey's theorem.)

If nG denotes n disjoint copies of a graph G and $n \geq 2$, then

$$R(nK_3) = R(nK_3, nK_3) = 5n.$$

If D is the four-vertex graph consisting of K_3 and an adjoined edge, and $n \geq 2$, then $R(nD) = 6n$.

CHAPTER 4

Extremal geometrical problems

This Chapter is devoted to yet another direction of extremal problems: concerning discrete families of geometrical objects. As an application we study the connection with extremal problems on hypergraphs, and also give some applications in matrix algebra.

1. Linear normed spaces

A nonempty set X is called a *linear space* over the set of real numbers $\mathbf{R}$ if an addition operation is given on X, with respect to which X is closed and has a neutral element 0, and if there is defined multiplication of elements of X by real numbers (from $\mathbf{R}$), with as result elements of X. In addition, the following conditions must hold: $\forall \alpha, \beta \in \mathbf{R}$, $x, y \in X$: $\alpha(\beta x) = (\alpha\beta)x$; $(\alpha+\beta)x = \alpha x + \beta x$; $\alpha(x+y) = \alpha x + \alpha y$.

A linear space is also called a *vector space*, and its elements $x \in X$ are also called *points* or *vectors*. Instead of over the set of real numbers $\mathbf{R}$, we may consider a linear space over the set of complex numbers $\mathbf{C}$ (or, in general, over an arbitrary field P).

A system of vectors $x_1, x_2, \ldots, x_d \in X$ of a linear space X is said to be *linearly dependent* if there are numbers $\alpha_1, \alpha_2, \ldots, \alpha_d \in \mathbf{R}$, not all zero, for which $\alpha_1 x_1 + \alpha_2 x_2 + \cdots + \alpha_d x_d = 0$. In the opposite case, i.e. such $\alpha_1, \alpha_2, \ldots, \alpha_d \in \mathbf{R}$ do not exist, this system of vectors is said to be *linearly independent*. We say that *the space X has dimension d* if it contains a system of d linearly independent vectors, while every system of $d+1$ vectors is linearly dependent. If there is a linearly independent system of d vectors for arbitrarily large d, the space is said to be *infinite dimensional*.

A linear space X is said to be *normed* if to each vector $x \in X$ there corresponds a number $\|x\|$, called the *norm* of this vector, such that the following conditions hold for $\alpha, \beta \in \mathbf{R}$, $x, y \in X$:

$$\|x\| \geq 0, \qquad x \neq 0 \qquad \text{(nonnegativity)},$$
$$\|\alpha x\| = |\alpha| \cdot \|x\| \qquad \text{(homogeneity)},$$
$$\|x + y\| \leq \|x\| + \|y\| \qquad \text{(convextity, or the triangle inequality)}.$$

A subset Y of a normed linear space X is called a *subspace* if it is itself a normed linear space with respect to the operations of addition of vectors and multiplication by scalars used in X. In other words, $Y \subset X$ is a subspace of X if $x, y \in Y$, $\alpha, \beta \in \mathbf{R}$ imply $\alpha x + \beta y \in Y$. The norm defined in X is also a norm in Y.

We consider some examples of linear normed spaces. Let $p \geq 1$. We denote by ℓ_p the space whose points are sequences of numbers $x = (x_1, x_2, \ldots)$ for which

$$\sum_{i=1}^{\infty} |x_i|^p < \infty.$$

The norm of such a sequence in ℓ_p is defined as the number

$$\|x\| = \left(\sum_{i=1}^{\infty} |x_i|^p \right)^{1/p}.$$

The triangle inequality for this norm is ensured by the condition $p \geq 1$. A space ℓ_p may be infinite dimensional, and may be finite dimensional, of dimension d. In the latter case its points are number sequences of length d (d-dimensional vectors), and for them the requirement that the sum of powers of their constituents is finite is clearly fulfilled.

A limit case of the spaces ℓ_p is the space of bounded number sequences with norm

$$\|x\| = \sup_i |x_i|.$$

We denote this space by ℓ_∞.

The infinite-dimensional space ℓ_2 is called the *Hilbert space*. A space l_2 of dimension $d < \infty$ is d-dimensional euclidean space, and is also denoted by $\mathbf{R}^d$. For any two vectors $x, y \in \ell_2$ we have the equality

$$\|x + y\|^2 + \|x - y\|^2 = 2(\|x\|^2 + \|y\|^2),$$

called the *parallelogram rule*, or *parallelogram law*. For ℓ_p with $p \neq 2$ this rule is not satisfied, in general, but we have a chain of inequalities:

$$2^{\min\{1, p-1\}} (\|x\|^p + \|y\|^p) \leq \|x + y\|^p + \|x - y\|^p \leq 2^{\max\{1, p-1\}} (\|x\|^p + \|y\|^p).$$

The space ℓ_2 has the very essential feature that in it a scalar product can be defined: for $x, y \in \ell_2$,

$$(x, y) = \tfrac{1}{4}(\|x + y\|^2 + \|x - y\|^2).$$

If $x = (x_1, x_2, \ldots)$, $y = (y_1, y_2, \ldots)$, then

$$(x, y) = \sum_i x_i y_i,$$

and $(x, x) = \|x\|^2$, $(x + z, y) = (x, y) + (z, y)$.

The *unit sphere* in a normed linear space X is the set of all $x \in X$ satisfying $\|x\| = 1$. The norm of each normed linear space is uniquely determined by the shape of its unit sphere. For example, in the two-dimensional case, the unit sphere in ℓ_2 is the circle of radius one; the unit sphere in ℓ_1 is the square with vertices $(0, 1)$, $(1, 0)$, $(0, -1)$, $(-1, 0)$; and the unit sphere in ℓ_∞ is the square with vertices $(1, 1)$, $(1, -1)$, $(-1, 1)$, $(-1, -1)$. In the three-dimensional case, the unit sphere in ℓ_2 is the ordinary three-dimensional sphere of radius one; the unit sphere in ℓ_1 is the tetrahedron with vertices at $(0, 0, 1)$, $(0, 1, 0)$, $(1, 0, 0)$, $(-1, 0, 0)$, $(0, -1, 0)$, $(0, 0, -1)$; and the unit sphere in ℓ_∞ is the cube with vertices at $(1, 1, 1)$, $(1, 1, -1)$, $(1, -1, 1)$, $(1, -1, -1)$, $(-1, 1, 1)$, $(-1, 1, -1)$, $(-1, -1, 1)$, $(-1, -1, -1)$.

We denote by $\sigma = \{x_1, x_2, \dots\}$ an unordered set (system) of vectors $x_i \in X$. When necessary, we indicate the number n of vectors in σ by a subscript: σ_n. The system σ may contain identical vectors, therefore the notation $\sigma_n \subset \sigma_m$, for $n \leq m$, means that if $\sigma_m = \{x_1, \dots, x_m\}$, then $\sigma_n = \{x_{i_1}, \dots, x_{i_n}\}$, where $1 \leq i_1 < \dots < i_n \leq m$. Hence there are $\binom{m}{n}$ possible choices of such n-subsystems. Sometimes we impose metrical restrictions on a system σ, of the type: $\sigma = \{x_1, x_2, \dots : \|x_i\| \geq 1, i = 1, 2, \dots\}$. These are always explicitly mentioned. We put $(\sigma) = \sum_{x \in \sigma} x$ and $\|\sigma\| = \|(\sigma)\| = \|\sum_{x \in \sigma} x\|$.

We note another useful identity. If $k \geq l \geq t$ and $\sigma_t \subset \sigma_k$, then

$$\sum_{\sigma_t \subseteq \sigma_l \subseteq \sigma_k} (\sigma_l) = (\sigma_t)\binom{k-t}{l-t} + (\sigma_k - \sigma_t)\binom{k-t-1}{l-t-1}. \tag{4.1}$$

In fact,

$$\sum_{\sigma_t \subseteq \sigma_l \subseteq \sigma_k} (\sigma_l) = \sum_{\sigma_t \subseteq \sigma_l \subseteq \sigma_k} \sum_{x \in \sigma_l} x =$$

$$= \sum_{\sigma_t \subseteq \sigma_l \subseteq \sigma_k} \sum_{x \in \sigma_k} x\chi\{x \in \sigma_l\} = \sum_{x \in \sigma_k} x \sum_{\sigma_t \subseteq \sigma_l \subseteq \sigma_k} \chi\{x \in \sigma_l\} =$$

$$= \sum_{x \in \sigma_t} x \sum_{\sigma_t \subseteq \sigma_l \subseteq \sigma_k} \chi\{x \in \sigma_l\} + \sum_{x \in \sigma_k - \sigma_t} x \sum_{\sigma_t \subseteq \sigma_l \subseteq \sigma_k} \chi\{x \in \sigma_l\} =$$

$$= (\sigma_t)\binom{k-t}{l-t} + (\sigma_k - \sigma_t)\binom{k-t-1}{l-t-1}.$$

2. Extremal geometric constants

The theme of extremal geometric constants includes both the calculation of extremal numerical characteristics of systems of vectors and the spatial description of systems of vectors that are extremal with respect to some property.

We will now compute several concrete extremal geometric constants.

CONSTANT A. Let $A(k, l; X)$ be the largest number A for which

$$\forall \sigma_k \subset X \ \exists \sigma_l \subset \sigma_k : \ \|\sigma_l\| \geq A\|\sigma_k - \sigma_l\|.$$

If $k \geq 2l$, then

$$A(k, l; X) = \frac{l}{k-l}; \tag{4.2}$$

if $k \leq 2l$, then

$$\inf_X A(k, l; X) = A(k, l; \ell_\infty) = \frac{l}{3l-k}. \tag{4.3}$$

PROOF. Let $\sigma'_l \subset \sigma_k$ be a subsystem such that

$$\max_{\sigma_l \subset \sigma_k} \|\sigma_l\| = \|\sigma'_l\|.$$

Then for $k \geq 2l$ we have by (4.1),

$$\sum_{\sigma_l \subseteq \sigma_k - \sigma'_l} (\sigma_l) = (\sigma_k - \sigma'_l)\binom{k-l-1}{l-1}.$$

Taking in this identity norms on both sides and using the triangle inequality, we obtain

$$\|\sigma'_l\| \geq \frac{l}{k-l}\|\sigma_k - \sigma'_l\|. \tag{4.4}$$

If $k \leq 2l$, then we have by (4.1),

$$\sum_{\sigma_k - \sigma'_l \subseteq \sigma_l \subseteq \sigma_k} (\sigma_l) = (\sigma_k - \sigma'_l)\binom{l}{2l-k} + (\sigma'_l)\binom{l-1}{2l-k-1},$$

and transition to norms gives

$$\|\sigma'_l\| \geq \frac{l}{3l-k}\|\sigma_k - \sigma'_l\|. \tag{4.5}$$

The fact that (4.4) and (4.5) cannot be improved upon in the class of all normed linear spaces is demonstrated by the following construct $\Sigma(k, l; \ell_\infty^k)$ of k unit vectors in ℓ_∞:

$$x_1 = (-1, (2l-1)^{-1}, \dots, (2l-1)^{-1}),$$
$$x_2 = ((2l-1)^{-1}, -1, \dots, (2l-1)^{-1}),$$
$$\dots\dots\dots$$
$$x_k = ((2l-1)^{-1}, (2l-1)^{-1}, \dots, -1).$$

Moreover, in each X equality in (4.4) is realized by a system of k vectors, which proves (4.2). $\square$

CONSTANT B. Let $B(k, r, l; X)$ be the largest number B for which

$$\forall \sigma_k \subset X \ \forall \sigma_r \subset \sigma_k \ \exists \sigma_l \subset \sigma_k : \ \|\sigma_l\| \geq B\|\sigma_r\|.$$

Then with the same methods it is easy to prove that

$$B(k, r, l; X) = \begin{cases} \frac{l}{r}, & k \geq r \geq l \geq 1, \\ \frac{l}{2l-r}, & k \geq l+r \geq 2r, \\ \frac{l(k-l)}{r(k+l-2r)}, & k \leq l+r, \quad l \geq r, \end{cases}$$

and there is a one-dimensional construct realizing these values.

PROOF. By (4.1), if $k \geq r \geq l \geq 1$ and $\sigma_r \subset \sigma_k$, then

$$\sum_{\sigma_l \subseteq \sigma_r} (\sigma_l) = (\sigma_r)\binom{r-1}{l-1},$$

hence we have

$$\max_{\sigma_l \subseteq \sigma_r} \|\sigma_l\| \geq \frac{l}{r}\|\sigma_r\|,$$

which clearly cannot be improved upon, as demonstrated by the pencil of k unit vectors. Let $k \geq l+r \geq 2r$, $\sigma_n \subset \sigma_k$. By (4.1) we have

$$\binom{k-r-1}{l-1}\sum_{\sigma_r \subseteq \sigma_l \subseteq \sigma_k} (\sigma_l) - \binom{k-r-1}{l-r-1}\sum_{\sigma_l \subseteq \sigma_k - \sigma_r} (\sigma_l) = (\sigma_r)\binom{k-1}{l-r}\binom{k-r-1}{l-1}.$$

This implies that $\|\sigma_l\| \geq l\|\sigma_r\|/(2l - r)$. Both these estimates cannot be improved upon in the class of unit vectors. It suffices to take for σ_k the system $\Sigma(k, l; \ell_\infty^k)$; moreover, the last inequality cannot be improved upon for all X.

Let $k \leq l + r$, $l \geq r$, $\sigma_r \subset \sigma_k$. Then by (4.1) we have

$$\left(\begin{matrix} r \\ l-k+r \end{matrix}\right) \sum_{\sigma_r \subseteq \sigma_l \subseteq \sigma_k} (\sigma_l) - \left(\begin{matrix} k-r-1 \\ l-r-1 \end{matrix}\right) \sum_{\sigma_k - \sigma_r \subseteq \sigma_l \subseteq \sigma_k} (\sigma_l) =$$

$$= (\sigma_r)\left(\left(\begin{matrix} k-1 \\ l-r \end{matrix}\right)\left(\begin{matrix} r \\ l-k+r-1 \end{matrix}\right) - \left(\begin{matrix} r-1 \\ l-k+r-1 \end{matrix}\right)\left(\begin{matrix} k-r-1 \\ l-r-1 \end{matrix}\right)\right),$$

whence $\|\sigma_l\| \geq l(k - l)\|\sigma_r\|/r(k + l - 2r)$. This estimate cannot be improved upon for all X. $\square$

CONSTANT C. Let $C(k, r, l; X)$ be the largest number C for which

$$\forall \sigma_k \subset X : \sum_{\sigma_r \subset \sigma_k} \|\sigma_r\| \geq C \sum_{\sigma_l \subset \sigma_k} \|\sigma_l\|.$$

Using the recurrence

$$C(k, r, l; X) \geq \prod_{i=r}^{l-1} C(k, i, i+1; X)$$

and the triangle inequality, we find

$$C(k, r, l; X) = \frac{\left(\begin{matrix} k-1 \\ r-1 \end{matrix}\right)}{\left(\begin{matrix} k-1 \\ l-1 \end{matrix}\right)}, \qquad l \geq r.$$

A system of k distinct vectors is an extremal construct in this case.

In general, if a system of vectors σ is free of any metrical restrictions, then the problem of computing the extremal geometric constant is essentially one-dimensional. In particular, this is demonstrated by the computations of the constants above. The situation changes if we introduce restrictions on the lengths of vectors.

CONSTANT delta. Let $\delta(l, k; X)$ be the largest number δ for which

$$\forall \sigma_k = \{x_1, \ldots, x_k : \|x_i\| \geq 1, \ i = 1, \ldots, k\} \subset X \ \exists \sigma_l \subset \sigma_k : \|\sigma_l\| \geq \delta.$$

In other words, we must compute the constant

$$\delta(l, k; X) = \inf_{\substack{\sigma_k \subset X \\ \|x_i\| \geq 1}} \max_{\sigma_l \subset \sigma_k} \|\sigma_l\|.$$

THEOREM 4.1. *Let X be a normed linear space, H a Hilbert space, and $\mathbf{R}^d$ d-dimensional euclidean space. Then the following formulas hold for $\delta(l, k; X)$:*

$$\inf_X \delta(l, k; X) = \delta(l, k; \ell_\infty) = \tfrac{l}{2l-1}, \qquad k > l; \tag{4.6}$$

$$\delta(l, l+1; \ell_1) = \tfrac{l}{2l-1}, \qquad \dim \ell_1 \geq l+1; \tag{4.7}$$

$$\forall k \ \exists d(k) : \ \delta(2, k; \ell_1) = \begin{cases} \frac{k-2}{k-1}, & k \text{ even, } \dim \ell_1 \geq d(k), \\ \frac{k-1}{k}, & k \text{ odd, } \dim \ell_1 \geq d(k); \end{cases} \tag{4.8}$$

moreover, $d(2) = 1$, $d(3) = d(4) = 3$, $d(5) = d(6) = 10$, $d(7) = d(8) = 7$;

$$\delta(2, k; \mathbf{R}^d) = 2^{0.5}, \qquad d + 2 \le k \le 2d; \tag{4.9}$$

$$\sup_{X:\dim X = \infty} \delta(l, l + 1; X) = \delta(l, l + 1; \ell_2); \tag{4.10}$$

$$\delta(l, k; H) = \left(\frac{l(k - l)}{k - 1}\right)^{0.5}, \qquad \dim H \ge k - 1; \tag{4.11}$$

moreover, (4.11) *is realized by the regular k-vertex simplex inscribed in the unit sphere of* $\mathbf{R}^{k-1}$;

$$\delta(k - 1, k; H) = 1, \qquad \dim H \ge 2, \tag{4.12}$$

$$\delta(2, k; \mathbf{R}^2) = 2 \cos \frac{\pi}{k}; \tag{4.13}$$

moreover, (4.12) *and* (4.13) *are realized by the system of vectors in the plane forming the vertices of the regular k-gon inscribed in the unit circle;*

$$\delta(3, k; \mathbf{R}^2) = \begin{cases} \left(5 + 4\cos\frac{2\pi}{\lfloor k/2 \rfloor}\right)^{0.5}, & k \ne 3, 5, \\ \frac{1 + 5^{0.5}}{2}, & k = 5 \\ 0, & k = 3; \end{cases} \tag{4.14}$$

if $\dim X = 1$, *then*

$$\delta(3, k; \mathbf{R}^2) = \begin{cases} l & k \ge 2l - 1 \\ k - l, & k < 2l - 1, \quad k \text{ even}, \\ \frac{l(k-l)}{l-1}, & k < 2l - 1, \quad k \text{ odd.} \end{cases} \tag{4.15}$$

We prove some of these formulas, marking, for convenience, the proof of these by their sequence numbers.

Put $k = l + 1$ in the A, then (4.6) follows immediately from (4.3). The definitions of B and *delta* imply the inequality

$$\delta(l, k; X) \ge B(k, r, l; X)\delta(r, k; X),$$

which for $r = 1$, $k > l$ also implies (4.6). Moreover, (4.6) is a direct consequence of Theorem 4.2 below.

To prove (4.7) it suffices to exhibit a system of vectors in ℓ_1^{l+1} with norm of a sum of any l of these equal to the righthand side of (4.7). Clearly, the following is such a system:

$$x_1 = \frac{(1 - l, 1, \dots, 1)}{2l - 1},$$

$$x_2 = \frac{(1, 1 - l, 1, \dots, 1)}{2l - 1},$$

$$\dots\dots\dots\dots$$

$$x_{l+1} = \frac{(1, \dots, 1, 1 - l)}{2l - 1},$$

since

$$\left\|\sum_{j=1}^{l+1} x_j - x_i\right\| = \frac{l}{2l-1}, \qquad i = 1, 2, \dots, l+1,$$

for it.

(4.15). The case $k \geq 2l - 1$ is trivial. Let $k < 2l - 1$. Consider k numbers, with absolute values not less than 1, and, for the sake of being specific, of which q are negative and p positive: $p + q = k$, $p \geq q$, $\{x_1, \dots, x_p, -y_1, \dots -y_q\}$, $1 \leq x_1 \leq \dots \leq x_p$, $1 \leq y_1 \leq \dots \leq y_q$. If $p \geq l$ or $q \geq l$, then there are l numbers of the same parity, and the absolute value of their sum is not less than l. Therefore we assume that $p, q < l$. It can be easily seen that the maximal absolute value of the sum of l distinct numbers from the k given numbers is taken by one of the following sums:

$$a_1 = x_1 + \dots + x_p - y_1 - \dots - y_{l-p}, \qquad a_2 = y_1 + \dots + y_q - x_1 - \dots - x_{l-q}.$$

However,

$$\max(|a_1|, |a_2|) \geq \left|\frac{qa_1 + (l-p)a_2}{l-p+q}\right| \geq \left|\frac{(pq - (l-p)(l-q))x_1}{l-p+q}\right| \geq \frac{l(k-l)}{l-p+q}.$$

It remains to note that the fraction $l(k-l)/(l-p+q)$ takes its smallest value for $p - q = 0$ if k is even, and for $p - q = 1$ if k is odd.

The constant δ can be viewed from a somewhat different perspective, allowing us to introduce and study some of its useful modifications. More precisely, with each concrete subsystem of vectors $\sigma_l \subset \sigma_k$ we may associate the sum

$$\sum_{i=1}^{k} \epsilon_i x_i,$$

where $\epsilon_i = \chi\{x_i \in \sigma_l\}$. So, we may associate to each selection of a concrete subsystem of vectors $\sigma_l \subset \sigma_k$ a rearrangement of k numbers ϵ_i, among which there are l ones and $k - l$ zeros. Hence, selecting a maximizing system $\sigma_l \subset \sigma_k$ is equivalent to mastering all rearrangements of these numbers, i.e.

$$\delta(l, k; X) = \inf_{\substack{\sigma_k \subset X \\ \|x_i\| \geq 1}} \max_{\pi \in S_k} \left\|\sum_{i=1}^{k} \epsilon_{\pi(i)} x_i\right\|,$$

where S_k is the symmetric group of all permutations on the set $[k] = \{1, 2, \dots, k\}$.

It is now clear that this mastering over all rearrangements can be done not only with vectorial coefficients, consisting of ones and zeros, but with arbitrary numerical vectors. Thus, we consider the following constant:

$$\delta(W; X) = \sum_{\substack{\sigma_k \subset X \\ \|x_i\| \geq 1}} \max_{\pi \in S_k} \left\|\sum_{i=1}^{k} w_{\pi(i)} x_i\right\|,$$

where $W = (w_1, \dots, w_k)$ is a fixed numerical vector. Hence, if W consists of l ones and $k - l$ zeros, the latter two constants coincide.

The constant $\delta(W; X)$ can be naturally called a *weighted geometric constant*. We put

$$(W, \pi, \sigma_k) = \sum_{i=1}^{k} w_{\pi(i)} x_i,$$

and consider the functional

$$\|W, \sigma_k\| = \max_{\pi} \|(W, \pi, \sigma_k)\|,$$

where the maximum is taken over all rearrangements $\pi \in S_k$. This functional is invariant under rearrangements of the components of W, hence in the sequel we assume that $w_1 \geq \cdots \geq w_k$ and $w = w_1 + \cdots + w_k$. The dimension of W is sometimes indicated by a subscript. The values of the weighted constant in the case of a degenerate vector, i.e. when there is a single component or no component at all, clearly are: $\delta(W_1; X) = |w|$, $\delta(0; X) = 0$. Hence in the sequel we consider only nondegenerate vectors. In general, the dependence on the shape of the weight vector is essential. For example, below we convince ourselves that a nontrivial part of the problem of computing the weighted constant is related to balanced vectors, i.e. those W for which for every i we have $w(w - w_i) > 0$. The least possible value of the weighted constant is given by:

THEOREM 4.2. *Let W_n be a nondegenerate weight vector. Then*

$$\inf_{X} \delta(W; X) = \delta(W_n; \ell_\infty) = \frac{|w(w_1 - w_n)|}{|w - w_1| + |w - w_n|}. \tag{4.16}$$

PROOF. Let $\pi_1, \ldots, \pi_k$ be rearrangements of the numbers $\{1, \ldots, n\}$, and $a_1, \ldots, a_k$ real numbers. Then

$$\sum_{j=1}^{k} |a_j| \|W_n, \sigma_n\| \geq \sum_{j=1}^{k} |a_j| \|W, \pi_j, \sigma_n\| \geq \left\| \sum_{a_j} (W_n, \pi_j, \sigma_n) \right\|,$$

so that

$$(|w - w_1| + |w - w_n|) \|W_n, \sigma_n\| \geq$$

$$\geq \frac{\left\| (w - w_n) \sum_{\pi: \pi(1)=1}(W_n, \pi, \sigma_n) - (w - w_1) \sum_{\pi: \pi(1)=n}(W_n, \pi, \sigma_n) \right\|}{(n-1)!} =$$

$$= \left\| (w - w_n) \left(w_1 x_1 + \frac{w - w_1}{n-1} \sum_{i=2}^{n} x_i \right) - (w - w_1) \left(w_n x_1 + \frac{w - w_n}{n-1} \sum_{i=2}^{n} x_i \right) \right\| =$$

$$= |w(w_1 - w_n)| \|x_1\| \geq |w(w_1 - w_n)|.$$

Thus, the lower bound in (4.16) has been proved.

If the weight vector is not balanced, then

$$\frac{|w(w_1 - w_n)|}{|w - w_1| + |w - w_n|} = |w|,$$

and (4.16) follows from the general upper bound $\delta(W; X) \leq |w|$, which follows for a system of n distinct vectors.

If the weight vector is balanced, then

$$\frac{|w(w_1 - w_n)|}{|w - w_1| + |w - w_n|} = \frac{w(w_1 - w_n)}{2w - w_1 - w_n}.$$

Put

$$a = \frac{w_1 + w_n}{2w - w_1 - w_n},$$

and consider in the space ℓ_∞ the system Σ_n of n vectors of the form

$$x_1 = (-1, a, \ldots, a),$$
$$x_2 = (a, -1, \ldots, a),$$
$$\cdots \cdots \cdots$$
$$x_n = (a, \ldots, a, -1).$$

Since W is balanced, we have $|w| > |w_1 + w_n|$, hence all vectors in Σ_n have norm one. Moreover, for any rearrangement π,

$$\|(W_n, \pi, \Sigma_n)\| = \max_{1 \le j \le n} |(w - w_{pi(j)})a - w_{\pi(j)}|,$$

and because the function $|(w - z)a - z|$ is convex in z, we have

$$\|(W_n, \pi, \Sigma_n)\| = \max(|(w - w_1)a - w_1|, |(w - w_n)a - w_n|) = \left| \frac{w(w_1 - w_n)}{2w - w_1 - w_n} \right|. \qquad \square$$

In the sequel it is convenient to introduce a special kind of vector W, namely: W has dimension m and $m - n$ zero components. We denote the final variant by

$$\delta(W_m, n; X) = \delta(W_m; X) = \inf_{\sigma_n, \sigma_m, X} \max_{\sigma_n \subset \sigma_m} \|W_m, \sigma_n\|.$$

A bound resembling the bound for the balanced case can be given also in the Hilbert situation.

THEOREM 4.3. *Let H be a Hilbert space of dimension at least $m - 1$, and let*

$$\sum_{\substack{i \neq j}}^{n} w_i w_j > 0.$$

Then

$$\delta(W_m, n; H) = \left(\frac{m \sum_{i=1}^{n} w_i^2 - w^2}{m - 1} \right)^{0.5}. \tag{4.17}$$

If $\dim H \ge 1$ *and*

$$\sum_{\substack{i \neq j}}^{n} w_i w_j \le 0,$$

then

$$\delta(W_m, n; H) = |w|.$$

PROOF. For

$$\sigma_m = \{x_1, \ldots, x_m : \|x_i\| \geq 1, \ i = 1, \ldots, m\} \subset H \tag{4.18}$$

we put $b_i = i\binom{m-i}{n-i}(n-i)!$,

$$A = \sum_{\sigma_n \subset \sigma_m} \sum_\pi (W_n, \pi, \sigma_n)^2,$$

where squaring is understood in the sense of the scalar product, and the inner summation is over all n-rearrangements. Then we have

$$A = b_1 \sum_{i=1}^n w_i^2 \sum_{i=1}^m x_i^2 + 2b_2 \sum i < j^n w_i w_j \sum_{i<j}^m x_i x_j =$$

$$= b_2 \sum_{i<j}^n w_i w_j \left(\sum_{i=1}^m x_i \right)^2 + \left(b_1 \sum_{i=1}^n w_i^2 - b_2 \sum_{i<j}^n w_i w_j \right) \sum_{i=1}^m x_i^2,$$

so that we have

$$|A| \geq m \left(b_1 \sum_{i=1}^n w_i^2 - b_2 \sum_{i<j}^n w_i w_j \right) \qquad \text{if } \sum w_i w_j > 0,$$

$$|A| \geq m^2 b_2 \sum_{i>j}^n w_i w_j + m \left(b_1 \sum_{i=1}^n w_i^2 - b_2 \sum_{i>j}^n w_i w_j \right) \qquad \text{if } \sum w_i w_j \leq 0.$$

Since the sum A contains $\binom{m}{n}n!$ summands, there is at least one such summand that is not less than the average of the two bounds given above, so that by elementary calculations (4.17) and (4.18) as lower bounds follow. Taking this into account, the fact that $\delta(W; X) \leq |w|$ immediately implies (4.18). To prove (4.17) it suffices to take as extremal construct the system of m vectors in $\mathbf{R}^{m-1}$ that are the vertices of a regular simplex $\Sigma_m(\mathbf{R}^{m-1})$ inscribed in the unit sphere of $\mathbf{R}^{m-1}$. For this construct all summands in A are equal, and $\|\Sigma_m(\mathbf{R}^{m-1})\| = 0$. So, equality holds in all estimates given above, and (4.17) is proved. $\square$

There is another modification of the constant δ which is useful. Here we give estimates of its value without proofs, since the proofs differ but slightly from those for the corresponding estimates for δ:

$$\delta_r(W; X) = \inf_{\substack{\sigma_n \subset X \\ \|x_i\| \geq 1 \\ i=1,\ldots,r}} \max_{\pi \in S_n} \left\| \sum_{i=1}^n w_{\pi(i)} x_i \right\|,$$

where $W = (w_1, \ldots, w_n)$ is a fixed numerical vector and S_n is the set of rearrangements of the indices $1, 2, \ldots, n$.

We put $w_{\max} = \max(w_1, \ldots, w_n)$, $w_{\min} = \min(w_1, \ldots, w_n)$,

$$\delta(W) = \frac{|w(w_{\max} - w_{\min})|}{|w - w_{\max}| + |w - w_{\min}|},$$

and consider nondegenerate vectors only, i.e. $W \neq 0$ and $n \geq 2$.

THEOREM 4.4. *The following formulas hold for* δ_r:

for every $r \leq n$,

$$\inf_X \delta_r(W; X) = \delta(W);$$

moreover, if W is not balanced, then $\forall X$: $\delta_r(W; X) = \delta(W) = |w|$, and if W is balanced and X contains a subspace isometric to ℓ_∞^r, then

$$\delta_r(W; X) = \delta(W);$$

for any space X and vector $W \in \mathbf{R}^n$,

$$\delta_1(W; X) = \delta(W);$$

$$\delta_r(W; \ell_2^k) \geq \begin{cases} \left(\frac{r}{n(n-1)}\left(n\sum_{i=1}^n w_i^2 - w^2\right)\right)^{0.5} & \text{if } \sum_{1 \leq i < j \leq n} w_i w_j \geq 0, \\ |w|\left(\frac{r}{n}\right)^{0.5} & \text{if } \sum_{1 \leq i < j \leq n} w_i w_j \leq 0. \end{cases}$$

Contact numbers. The constant δ is tightly related to the Newton—Gregory contact numbers (the largest number of unit spheres that simultaneously can touch a central unit sphere). Indeed, if we consider the extremal constant

$$d(k; X) = \sup_{\substack{\sigma_k \subset X \\ \|x_i\|=1}} \min_{1 \leq i < j \leq n} \|x_i - x_j\|,$$

then the contact number $k(X)$ (for a space X) can be defined as the largest integer k satisfying the inequality $d(k; X) \geq 1$. If X is a Hilbert pace, then the parallelogram law holds in it: the sum of the squares of the sides of a parallelogram is equal to the sum of the squares of its diagonals. Hence, if X is a Hilbert space, then $d^2(k; X) + \delta^2(2, k; X) = 4$, and hence the contact number $k(X)$ is the largest integer k satisfying $\delta^2(2, k; X) \leq 3$.

In a euclidean space the exact values of contact numbers have been computed for a few dimensions only: $k(\mathbf{R}^2) = 6$, $k(\mathbf{R}^3) = 12$, $k(\mathbf{R}^8) = 240$, $k(\mathbf{R}^{24}) = 196\,560$. The dependence of the values of contact numbers on the norm of the space (in other words, on the shape of the unit sphere in it) is very important. It is easy to construct an example of a space for which the contact number in dimension three equals 26. Clearly, this value is realized in the three-dimensional space whose unit sphere has the shape of the ordinary three-dimensional cube, i.e. in ℓ_∞. In this case the system of unit spheres realizing the contact number 26 is represented by three layers of unit cubes, with nine cubes in each layer (located precisely as the smaller cubes are in Rubik's cube). In this case the central (invisible) cube touches all other cubes, while it touches certain cubes along faces, others along edges, and still others at vertices only. But let us return to euclidean space. Let $M_d(r)$ be the largest number of unit vectors in $\mathbf{R}^d$ whose pairwise distances are at least r, and let $N_d(s)$ be the largest integer N for which there are vectors $x_1, \ldots, x_N \in \mathbf{R}^d$ satisfying the conditions

$$(x_i, x_i) = 1, \qquad i = 1, \ldots, N,$$
$$(x_i, x_j) \leq s, \qquad i \neq j.$$

If $s = 1 - r^2/2$, then $M_d(r) = N_d(s)$, and $N_d(1/2) = k(\mathbf{R}^d)$.

To estimate $N_d(s)$ we use the method of spherical polynomials.

Consider the polynomials $Q_i^{(d)}(t)$ recursively defined by the relations

$$Q_0^{(d)}(t) = 1, \qquad Q_1^{(d)}(t) = dt,$$

$$Q_k^{(d)}(t) = \frac{d + 2k - 2}{k} \left(t Q_{k-1}^{(d)}(t) - \frac{d + k - 4}{d + 2k - 6} Q_{k-2}^{(d)}(t) \right), \qquad k \geq 2.$$

There is a map

$$f_k^{(d)} : \mathbf{R}^d \to \mathbf{R}^b, \qquad b = \binom{k + d - 2}{d - 2} + \binom{k + d - 3}{d - 2},$$

having the following properties. For any two vectors $x, y \in \mathbf{R}^d$, where $(x, x) = (y, y) = 1$, the following equation holds:

$$Q_k^{(d)}((x, y)) = (f_k^{(d)}(x), f_k^{(d)}(y)).$$

In particular, this implies that for any vectors $x_1, \ldots, x_N \in \mathbf{R}^d$ with $(x_i, x_i) = 1$, $i = 1, \ldots, N$:

$$\sum_{i,j=1}^{N} Q_k^{(d)}((x_i, x_j)) = \sum_{i,j=1}^{N} (f_k^{(d)}(x_i), f_k^{(d)}(x_j)) = \left(\sum_{i=1}^{N} f_k^{(d)}(x_i), \sum_{i=1}^{N} f_k^{(d)}(x_i) \right) \geq 0.$$

Consider now an arbitrary function

$$h(t) = \sum_k \lambda_k Q_k^{(d)}(t),$$

where $\lambda_k \geq 0$ for $k = 1, 2, \ldots$, and $h(t) \leq 0$ for $-1 \leq t \leq s$. If the vectors $x_1, \ldots, x_N \in \mathbf{R}^d$ satisfy the conditions for the constant $N_d(s)$, then

$$Nh(1) = \sum_{i=1}^{N} h((x_i, x_i)) \geq \sum_{i=1}^{N} h((x_i, x_i)) + 2 \sum_{1 \leq i < j \leq N} h((x_i, x_j)) = \sum_{i,j=1}^{N} h((x_i, x_j)) =$$

$$= \sum_{i,j=1}^{N} \sum_k \lambda_k Q_k^{(d)}((x_i, x_j)) = \sum_k \lambda_k \sum_{i,j=1}^{N} Q_k^{(d)}((x_i, x_i)) \geq \lambda_0 \sum_{i,j=1}^{N} Q_0^{(d)}((x_i, x_j)) = \lambda_0 N^2,$$

so that

$$N \leq \frac{h(1)}{\lambda_0}.$$

By constructing a suitable function h we can thus obtain an upper bound for the constant $N_d(s)$.

If, for $d = 8$, we consider the function

$$h(t) = (t + 1)\left(t + \tfrac{1}{2}\right)^2 t^2 \left(t - \tfrac{1}{2}\right) =$$

$$= \tfrac{9}{160} \left(\tfrac{1}{6} Q_0^{(8)} + \tfrac{2}{7} Q_1^{(8)} + \tfrac{20}{63} Q_2^{(8)} + \tfrac{208}{693} Q_3^{(8)} + \tfrac{808}{3003} Q_4^{(8)} + \tfrac{320}{3003} Q_5^{(8)} + \tfrac{512}{41769} Q_6^{(8)} \right),$$

then we immediately obtain

$$N_8 \left(\tfrac{1}{2} \right) \leq \frac{h(1)}{\lambda_0} = \frac{h(1)960}{9} = 240.$$

If, for $d = 24$, we take for h the function

$$h(t) = (t + 1)\left(t + \tfrac{1}{2}\right)^2 \left(t + \tfrac{1}{4}\right)^2 t^2 \left(t - \tfrac{1}{4}\right)^2 \left(t - \tfrac{1}{2}\right),$$

then we obtain by similar transformations:

$$N_{24}\left(\frac{1}{2}\right) \leq 196\,560.$$

d	$k(\mathbf{R}^d)$	d	$k(\mathbf{R}^d)$	d	$k(\mathbf{R}^d)$
1	2	9	306–380	17	5346–12 215
2	6	10	500–595	18	7398–17 877
3	12	11	582–915	19	10 668–25 901
4	24-25	12	840–1416	20	17 400–37 974
5	40–46	13	1130–2233	21	27 720–56 852
6	72–82	14	1582–3492	22	49 896–86 537
7	126–140	15	2564–5431	23	93 150–128 096
8	240	16	4320–8313	24	196 560

Table 4.1

There are known (already for thirty years) concrete placements of spheres realizing these bounds, and giving precise values of contact numbers in euclidean spaces of the indicated dimensions. The exact values and best bounds for contact numbers have been given in Table 4.1.

3. Some applications of geometric constants

It is remarkable that in its original form, Mantel's theorem was formulated and proved not in terms of graphs, but for vectors in euclidean space. An essential step in unifying extremal geometric problems and problems concerning graphs was made by P. Turan, who noted that among n vectors in a euclidean space there must be 'many' pairs of such vectors with 'long' sums. A sufficiently general revelation of this connection is given by

THEOREM 4.5. *Let $n \geq k \geq l \geq 1$ be natural numbers, and let X be a normed linear space. Then there are in each system $\sigma_n = \{x_1, \ldots, x_n: \|x_i\| \geq 1, \ i = 1, \ldots, n\} \subset X$ at least $T(n, k, l)$ subsystems $\sigma_l \subset \sigma_n$ such that*

$$\|\sigma_l\| \geq \delta(l, k; X).$$

PROOF. Taking the system $\sigma_n \subset X$ as vertex set, we construct an l-homogeneous hypergraph $G^l(\sigma_n)$ by the rule:

$$\sigma_l \in G^l(\sigma_n) \Leftrightarrow \|\sigma_l\| \geq \delta(l, k; X),$$

and we verify that this hypergraph has at least $T(n, k, l)$ hyperedges. Assume the contrary; then

$$\exists \sigma_k^1 \subset \sigma_n : \forall \sigma_l \subset \sigma_k^1 : \|\sigma_l\| < \delta(l, k; X),$$

or

$$\exists \sigma_k^1 \subset \sigma_n : \max_{\sigma_l \subset \sigma_k^1} \|\sigma_l\| < \delta(l, k; X).$$

But then

$$\min_{\sigma_k \subset X} \max_{\sigma_l \subset \sigma_k} \|\sigma_l\| \leq \max_{\sigma_l \subset \sigma_k^1} \|\sigma_l\| < \delta(l, k; X),$$

so that

$$\min_{\sigma_k \subset X} \max_{\sigma_l \subset \sigma_k} \|\sigma_l\| < \delta(l, k; X),$$

contradicting the definition of $\delta(k, l; X)$. □

A probabilistic meaning of the relation between geometric and combinatorial extremal constants was given by G. Katona. Its essence is that if among n vectors in a euclidean space there are 'many' pairs of vectors with 'long' sums, then by randomly choosing pairs of vectors, the probability that a pair has 'long' sum is large. An explicit probabilistic relation between combinatorial and geometric constants is demonstrated by

COROLLARY 4.1. *Let* $n \geq k \geq l \geq 1$ *be natural numbers,* X *a normed linear space, and* $\xi_1, \ldots, \xi_l$ *independent, identically distributed random vectors in* X. *Let* $T(n, k, l)$ *be Turán's number and* $\delta(l, k; X)$ *the geometric constant defined in §4.2. Then the following inequality holds:*

$$\mathcal{P}\left\{\left\|\sum_{i=1}^{l} \xi_i\right\| \geq x\delta(l, k; X)\right\} \geq l! \left(\lim_{n \to \infty} \frac{T(n, k, l)}{n^l}\right) \mathcal{P}^l\{\|\xi_i\| \geq x\}.$$

For instance, if X is an arbitrary linear normed space, $k = 3$ and $l = 2$, then

$$\mathcal{P}\left\{\|\xi + \eta\| \geq \frac{2}{3}x\right\} \geq \frac{1}{2}\mathcal{P}^2\{\|\xi\| \geq x\}; \tag{4.19}$$

however, if $X = H$, then

$$\mathcal{P}\{|\xi + \eta| \geq x\} \geq \frac{1}{2}\mathcal{P}^2\{|\xi| \geq x\}. \tag{4.20}$$

We give an application of (4.20) to the *Bertrand paradox*: what is the probability that the length of a 'random chord' in the unit disk is larger than $\sqrt{3}$? The essence of the Bertrand paradox is that there are several natural values for this probability, in particular $1/2$, $1/3$, $1/4$ (see [207]).

On the other hand, for independent, identically distributed random points in the unit disk,

$$\mathcal{P}\{|\xi - \eta| < \sqrt{3}\} = \mathcal{P}\{|\xi + \eta| < 1\} = 1 - \mathcal{P}\{|\xi + \eta| \geq 1\} \leq (4.20) \leq$$

$$\leq 1 - \frac{1}{2}\mathcal{P}^2\{|\xi| \geq 1\} = 1 - \frac{1}{2} = \frac{1}{2}.$$

Therefore, if both ends of the chord are independent and identically distributed, then the maximum of the probability considered in the Bertrand paradox is $1/2$.

More generally, if S is a bounded subset of $\mathbf{R}^n$ and ξ, η are independent and identically distributed, then the maximum probability of the event $|\xi - \eta| > a$ is equal to $(k - 1)/k$, where k is the maximum number of points $x_1, \ldots, x_k \in S$ such that $|x_i - x_j| > a$, $i \neq j$.

We now note some applications of geometric constants to matrices. We consider square matrices of order n. The classical notions in matrix algebra are: matrix norm, generalized matrix norm, spectral norm, and numerical radius.

A *generalized matrix norm* is a numerical function N on matrices satisfying the following conditions

$$N(A) \geq 0;$$

$N(A) > 0$ if $A \neq 0$;

$N(\alpha A) = |\alpha| N(A)$ for $\alpha \in \mathbf{C}$;

$N(A + B) \leq N(A) + N(B)$.

If, moreover, N satisfies $N(AB) \leq N(A)N(B)$, then it is called a *matrix norm*. An example of a matrix norm is the *spectral norm*

$$\|A\|_2 = \max\{x^T A^T A x : x \in \mathbf{C}^n, \|x^T\| = \|x\| = 1\}.$$

An example of a generalized matrix norm (which is not a matrix norm) is the *numerical radius*

$$r(A) = \max\{|x^T A x| : x \in \mathbf{C}^n, \|x^T\| = \|x\| = 1\}.$$

Consider a matrix C such that $C \neq \lambda I$ and $\operatorname{Tr} C \neq 0$ (here, $\operatorname{Tr} C$ denotes the trace of the matrix C). We define a generalized matrix norm r_C by

$$r_C(A) = \max\{|\operatorname{Tr}(CU^T AU)| : U^T = U = I\}.$$

If $C = \operatorname{diag}\{1, 0, \ldots, 0\}$, then $r_C(A)$ is the numerical radius of A.

For any generalized matrix norm N there exists a number $\nu(N)$, the smallest multiplier ν for which νN is a matrix norm (if $\nu(N) \leq 1$, then N is already a matrix norm). A matrix with real entries is called *Hermitean* if all its eigenvalues are real.[1]

THEOREM 4.6. *Let C be a Hermitean matrix and let $W = (w_1, \ldots, w_n)$ be the vector of eigenvalues of C (i.e. w_j is an eigenvalue of C). If $\delta_1(W; \mathbf{R}) \neq 0$, then*

$$\nu(r_C) \leq \frac{4|w_1 + \cdots + w_n|}{\delta_1^2(W; \mathbf{R})}.$$

To prove this theorem we need several lemmas.

LEMMA 4.1. *Let N be a generalized matrix norm, M a matrix norm, and $b \geq a > 0$ constants such that for any matrix A,*

$$aM(A) \leq N(A) \leq bM(A).$$

Then $\nu(N) \leq ba^{-2}$.

PROOF. Put $N'(A) = ba^{-2}N(A)$. Then

$$N'(AB) = ba^{-2}N(AB) \leq ba^{-2}M(AB) \leq ba^{-2}M(A)M(B) \leq$$
$$\leq ba^{-4}N(A)N(B) = N'(A)N'(B). \qquad \square$$

---

[1] Recall that a number x_i is an *eigenvalue* of a square matrix A if and only if

$$\det(x_i I - A) = 0.$$

LEMMA 4.2. *Let A and C be normal matrices[2] with eigenvalues $x_1, \ldots, x_n$ and $w_1, \ldots, w_n$, respectively. Then*

$$r_C(A) = \max_{\pi \in S_n} \left| \sum_{j=1}^{n} w_{\pi(j)} x_j \right|,$$

$$r_C(A) \leq \left| \sum_{j=1}^{n} w_j \right| \|A\|_2.$$

PROOF OF THE THEOREM. Let H be a Hermitean matrix with eigenvalues $y_1, \ldots, y_n$. Then

$$\|H\|_2 = \max_j |y_j|,$$

and, by Lemma 4.2,

$$r_C(H) = \max_{\pi \in S_n} \left| \sum_{j=1}^{n} w_{\pi(j)} y_j \right| \geq \delta(W; \mathbf{R}) \max_j |y_j| = \delta(W; \mathbf{R}) \|H\|_2.$$

It can be easily seen that $r_C(A) = r_C(A^T)$. Put $H_1 = A + A^T$, $H_2 = i(A - A^T)$. Then $A = (H_1 - iH_2)/2$, and the matrices H_1, H_2 are Hermitean. Since r_C is a generalized matrix norm, we have

$$r_C(A + A^T) \leq 2r_C(A), \qquad r_C(iA - iA^T) \leq 2r_C(A).$$

Therefore

$$r_C(A) \geq \frac{r_C(H_1) + R_C(H_2)}{4} \geq \delta(W; \mathbf{R}) \frac{\|H_1\|_2 + \|H_2\|_2}{4} \geq$$

$$\geq \delta(W; \mathbf{R}) \frac{\|H_1\|_2 - i\|H_2\|_2}{4} = \delta(W; \mathbf{R}) \frac{\|A\|_2}{2}.$$

Whence

$$\delta(W; \mathbf{R}) \frac{\|A\|_2}{2} \leq r_C(A) \leq \left| \sum_{j=1}^{n} w_j \right| \|A\|_2;$$

and by applying Lemma 4.1 we arrive at the required bound. □

COROLLARY 4.2. *Let C be a Hermitean matrix and $w_1 \geq w_2 \geq \cdots \geq w_n$ its eigenvalues, $w = w_1 + \cdots + w_n$, $w_1 \neq w_n$, $w \neq 0$. Then*

$$\nu(r_C) \leq \frac{4(|w - w_1| + |w - w_n|)}{|w|(w_1 - w_n)^2}.$$

As an example we note that for the classical numerical radius (the case $C = \{1, 0, \ldots, 0\}$) this corollary gives a sharp bound, since $\nu(r) = 4$.

COROLLARY 4.3. *Under the conditions of theorem 4.6,*

$$\delta(W) = \frac{|w(w_{max} - w_{min})|}{|w - w_{max}| + |w - w_{min}|} \geq 2|w|^{0.5},$$

where r_C is a matrix norm.

---

[2] A square matrix A with real entries is *normal* if $AA^T = A^T A$.

A square matrix of order n and with entries 1 and -1 is called a *Hadamard matrix* if it satisfies $AA^T = nI$. The basic problem concerning Hadamard matrices reduces to the question of existence: For which n is there a Hadamard matrix of order n? It is not difficult to verify that Hadamard matrices can exist only if $n = 1, 2, 4k$. The following result, which we state without proof, reduces this problem to the problem of computing an extremal geometric constant.

THEOREM 4.7. *The following assertions are equivalent:*

1) $\delta(2, 4n - 1; \ell_1^{4n-1}) = \frac{4n-2}{4n-1}$;
2) $\delta(2, 4n; \ell_1^{4n-1}) = \frac{4n-2}{4n-1}$;
3) *there exists a Hadamard matrix of order $4n$.*

4. Problems and assertions

PROBLEM 4.1. Does there exist a three-dimensional normed linear space with contact number of its unit sphere exceeding 26?

PROBLEM 4.2. If the weight vector is not balanced, then for all X,

$$\delta(W_m, m; X) = |w|.$$

PROBLEM 4.3. If the weight vector is of definite sign, and $|w_1| \geq \cdots \geq |w_n| > 0$, $m > n$, then

$$\inf_X \delta(W_n, m; X) = \delta(W_n, m; \ell_\infty) = \frac{ww_1}{|2w - w_1|}.$$

PROBLEM 4.4. If the weight vector is balanced and of varying sign, then

$$\inf_X \delta(W_n, m; X) = \delta(W_n, m; \ell_\infty) = \frac{w(w_1 - w_n)}{2w - w_1 - w_n}.$$

PROBLEM 4.5. If the weight vector is balanced and $w_1 = -w_n$, then

$$\inf_X \delta(W_n, m; X) = \delta(W_n, m; \ell_\infty) = w_1,$$

and, moreover, a system of m basis vectors in ℓ_∞ is an extremal construct.

PROBLEM 4.6. If Σ_k is a simplex inscribed in the unit sphere of $\mathbf{R}^{k-1}$ and $w_1, \ldots, w_i$ are real numbers, then for each $\sigma_l = \{x_1, \ldots, x_l\} \subset \Sigma_k$ we have

$$\left|\sum_{i=1}^{l} w_i x_i\right| = \left(\frac{k \sum_{i=1}^{l} w_i^2 - w^2}{k - 1}\right)^{0.5};$$

in particular, the length of an edge of Σ_k equals $(2k/(k-1))^{0.5}$.

PROBLEM 4.7. If $\dim H \geq n - 1$, $E_n = (\epsilon_1, \ldots, \epsilon_n)$, $\epsilon_i \in \{+1, -1\}$, $\epsilon^2 \geq n$, then

$$\delta(E_n; H) = \left(\frac{n^2 - \epsilon^2}{n - 1}\right)^{0.5}.$$

If, however, $\dim H \geq 1$, $\epsilon^2 \leq n$, then $\delta(E_n; H) = |\epsilon|$.

PROBLEM 4.8. If $\dim H \geq n - 1$, $E_n = (\epsilon_1, \ldots, \epsilon_n)$, $\epsilon_i \in \{+1, -1\}$, then

$$
\sup_{E_n} \delta(E_n; H) = \begin{cases}
1, & n = 1, 3, \\[2mm]
\left(\dfrac{n^2 - \lceil n^{0.5} \rceil^2}{n-1}\right)^{0.5}, & n -]n^{0.5}[\text{ even,} \quad n \neq 1, \\[3mm]
\left(\dfrac{n^2 - (\lceil n^{0.5} \rceil + 1)^2}{n-1}\right)^{0.5}, & n -]n^{0.5}[\text{ odd,} \quad n \neq 3.
\end{cases}
$$

PROBLEM 4.9. Define the constant $\delta(2, k_1, \ldots, k_t; X)$ as follows:

$$
\delta(2, k_1, \ldots, k_t; X) = \inf_{\sigma_k \subset X} \max_{\sigma_2 \subset \sigma_{k_i}} \min \|\sigma_2\|,
$$

where the maximum is over all partitions of the system of vectors σ_k, as a multiset, into blocks of prescribed dimensions:

$$
\sigma_k = \sum_{i=1}^{t} \sigma_{k_i},
$$

and the minimum is over only those blocks that have volume at least two. Then every system $\sigma_n = \{x_1, \ldots, x_n \colon \|x_i\| \geq 1, \, i = 1, \ldots, n\} \subset X$ contains at least

$$
m\left(n; \sum_{i=1}^{t} K_{k_i}\right)
$$

subsystems $\sigma_2 \subset \sigma_n$ such that $\|\sigma_2\| \geq \delta(2, k_1, \ldots, k_t; X)$. For the definition of $m(n; H_k)$ see Chapter 3.

PROBLEM 4.10. Try to compute or estimate the value of $\delta(2, k_1, \ldots, k_t; X)$.

PROBLEM 4.11. A system of unit vectors σ_n is called an *l-system* if

$$
\forall \sigma_p \subset \sigma_n \, \exists \sigma_q \subset \sigma_p \, \forall \sigma_k \subset \sigma - q : \, \|\sigma_k\| = l \in \mathbf{R}.
$$

Compute or estimate the constant

$$
\Delta_c(n, p, q, k; X) = \sup \|\sigma_n\|,
$$

where the supremum is over all *l-systems* σ_n.

Hint. Use the results in Chapter 3 concerning locally Turán hypergraphs.

PROBLEM 4.12. Suppose six points in the plane satisfy the inequalities $|a_i - a_j| \leq 1$ $(1 \leq i < j \leq 6)$. Prove that there are points a_k, a_l, a_m among these six points such that $|a_k - a_l| < 1$, $|a_k - a_m| < 1$, $|a_l - a_m| < 1$.

Hint. Use Ramsey's theorem.

CHAPTER 5

Application of the results of solving extremal combinatorial problems

The efficiency of the functioning of automatic control systems is directly related to an increase in the scientific–technological level of designing them. However, up till now the solution of a number of problems in the technical design of automatic control systems is mainly based on attempts by and intuition of the developers. Such an empirical approach can lead already in various stages of development of automatic control systems to errors in design solutions, the removal of the majority of which can only be managed at a very high price. This relates in full measure to the solution of the problem of choosing the necessary size of operation computer memory, to the solution of the problem of optimal splitting of programs and variables, and to a number of other problems arising in the process of designing automatic control systems.

The main instrument of study in the process of designing automatic control systems is imitation modeling. However, its use comes to a long path, and requires a large amount of time [96]. On the one hand, using this method for designing automatic control systems ensures the necessary precision in estimates of values of the parameters investigated, at each stage of the design, and on the other hand it practically leads to a halt at the stage of determining the parameters of a number of technical tools from the general flow of development. For example, nowadays the final choice of the necessary size of operation computer memory is successful only at the stage of test run of the automatic control system [40].

A new theoretical approach to the solution of the problem of designing an automatic control system is the use of methods from combinatorial analysis, and, in particular, from the theme of extremal combinatorial problems concerning number partitions. The high level of abstraction of the formulation and solution of extremal combinatorial problems allows us to use them in the design of both hardware and software of automatic control systems. Combinatorial methods of study assume a formalization of the various elements of a system by combinatorial objects. The collection of these objects forms combinatorial models, which make possible, on the basis of a priori information on the functioning of the elements in the system, the description of the whole set of their states. The use of the results of solving extremal combinatorial problems in the process of investigation essentially shortens the necessary amount of system states to be analyzed, and allows us to give a comparative analysis of indices of functioning, both using exact values of them and using (upper or lower) bounds of their values.

The principal aim of the examples of using the theme of extremal combinatorial problems concerning the set of number partitions lies in that the reader develops methodical experience on the formalization of the processes investigated here, using the notion of packability of partitions and extremal results for solving practical problems. Therefore, in the statements of the specific practical problems we do not give precise definitions of the processes investigated, and do not describe their cause–effect relationships with the process of functioning of an automatic control system in the large.

1. Combinatorial models for studying the process of computer memory allocation of an automatic control system

Studies related to improving the efficiency of methods of controlling computer memory allocation are mainly directed at the search for efficient methods for allocation, re-allocation, and reorganization of memory. *Allocation* of memory is a finite sequence of maps $(I \to F)_t$ $(t = 1, \dots)$ from a set I of information objects (programs, array data) or their names into the set F of physical addresses of the memory to be allocated at discrete time moments t of functioning of the automatic control system. *Re-allocation* of computer memory is the transfer of a number of information objects from the address space in operation memory into auxiliary memory with the aim of emptying operation memory and distributing in it other information objects necessary for continuing the computing process. A *reorganization* of memory is a displacement of information objects in address space. Re-allocation and reorganization of memory are one of the basic methods for improving the efficiency of using this costly computing resource of modern computers.

There are two methods of allocating memory: statical and dynamical. A memory allocation is called *static* if $(I \to F)_t$ is chosen once and for all for executing the program of problem. In *dynamic* memory allocation each $(I \to F)_t$ is chosen in the course of the computing process immediately at time t, starting from $(I \to F)_{t-1}$. The application of one or the other method depends on the presence of information:

- on memory usage;
- on properties of program referencing or the sequential use of information.

Static allocation can only be applied if information on memory usage and program referencing are available before solution by the program. The use of dynamic memory allocation assumes that information on usage is not known in advance, and that the property of referencing is determined during the execution of the program only. It is this method of functioning that is present in real-time automatic control systems. In such systems the requirements on operation memory are, in each concrete instance, determined by the nature and intensity of the flow of data for processing information, which, in turn, is random. In [14] it has been noted that efficient functioning of a real-time automatic control system can only be achieved when for the satisfaction of a demand to distinguish operation memory fewest possible restrictions are imposed, while emptying of occupied memory sectors is done as fast as possible. These thoughts speak in favor of dynamic allocation of operation computer memory.

Studies related to estimating the efficiency of applying various methods for controlling memory allocation pursuit mainly the following unique goals:

- free the programmer from worries about memory allocation;
- improve the efficiency of memory usage;
- minimize the expense of processing time in controlling memory allocation.

In realizing both the static and the dynamic means of memory allocation, one of the main obstacles on the way of an efficient use of them is memory fragmentation [41], [95]. In studies of the appearance of memory fragmentation we can distinguish two directions: *stochastic*, when the influence of fragmentation is considered as a random process, and *deformational*, when the process of functioning of the system leads to given, designed states of fragmented memory. The first of these approaches is related to studies of the process of memory allocation in computers with segmented organization of programs and data, the second to studies of stackwise memory organization or to memory allocation by size-restricted free sectors.

The loss in efficiency when using memory with segmented organization of programs and data is conditioned by the influence of *external fragmentation* or 'scattering' of memory at practically every time instant over the large amount of free and occupied sectors of various lengths. External fragmentation arises because of the random nature of the flow of inquiries on memory isolation, of the various sizes of these inquiries, which in the address memory space are placed like words, and also because of the random sojourn time of programs and data in computer memory.

'Scattering' of memory in the process of functioning of an automatic control system quite often leads to situations when in memory there is no free continuous sector of address space necessary to fulfill an entering memory inquiry. In this case, even if the total size of all available free fragments is equal or greater than the size of the part of free memory required, the entering inquiry cannot be satisfied without applying reorganization or re-allocation of memory. The very application of reorganization or re-allocation of memory requires an additional expense of the processing time in controlling memory allocation, which results in a reduction of the productivity of the automatic control system in the large. The loss in efficiency of memory usage when it is stackwise organized is conditioned by the influence of *internal fragmentation*. Internal fragmentation arises because of rounding the size of each entering memory inquiry to an integer number of stacks. This additionally distinguished sector of memory is not used in the process of executing the program, and also determines a loss in the efficient use of memory in the large.

Stackwise memory organization essentially simplifies the solution of the problem of memory allocation, since the size of every stack is the same, and for every concretely chosen stack we may take any other stack. However, studies indicate that in the process of functioning of an automatic control system the loss in efficiency of memory usage conditioned by the influence of internal fragmentation turns out to be larger than the loss brought about by the influence of external fragmentation [41], [173]. Hence, by reducing the expenses in processing time for controlling memory allocation for segmented organization of programs and data, this can also greatly increase the efficiency of such a mechanism of memory control. Therefore our investigations will be directed to the study of processes of memory allocation of computing systems with segmented organization of programs and data, which have the following advantages in comparison with systems with stackwise memory organization:

- the solution of the problem of organizing exterior referencing to segments is

greatly simplified, since in this case the unified programs are not required to work with absolute addresses;
- the control is made easier by the solution of re-entering programs;
- loss in efficiency in memory usage because of rounding the sizes of inquiries to the stack size adopted in the system (loss by internal fragmentation) is excluded.

We consider several general combinatorial models allowing the study of the process of allocation of operation computer memory when there is segmented organization of programs and data.

MODEL 5.1. At any time instant of functioning of the automatic control system, the influence of external fragmentation on the process of memory allocation can be sufficiently well characterized by the following parameters:
- the amount of free (occupied) sectors of memory;
- the size of the free (occupied) sectors;
- the total size of free (occupied) memory.

In these studies, inquiries on distinguishing memory are, at any time instant of functioning of the automatic control system, sufficiently well determined by the following parameters:
- the amount of inquiries on distinguishing memory in queue;
- the required sizes of continuous sectors of address memory space, or the sizes of the inquiries;
- the total size of memory required to satisfy the inquiries from the queue.

Let Q be the size of operation computer memory in an automatic control system, and N the total size of free memory, which takes values $N \in \mathbf{Z}^+$, $N \leq Q$, in the course of functioning of the system, where $\mathbf{Z}^+$ is the set of nonnegative integers. Because of the influence of external fragmentation of memory, n is 'scattered' into r free fragments, represented by continuous sectors of address memory space. Such a state of free memory may be interpreted as a vector

$$z(N) = (n_1, \ldots, n_r), \qquad N = \sum_{i=1}^{r} n_i, \qquad n_1 \geq n_2 \geq \cdots \geq n_r,$$

$n_i \in \mathbf{Z}^+$, where n_i is the size of the ith free sector of memory and r is the amount of such sectors.

DEFINITION 5.1. Two states of free memory $z(N) = (n_1, n_2, \ldots, n_r)$ and $z'(N) = (n'_1, n'_2, \ldots, n'_r)$ are regarded as distinct if they are different as vectors, i.e. if there is an i such that $n_i \neq n'_i$.

We can similarly interpret each state of occupied memory by a vector $g(D) = (d_1, d_2, \ldots, d_l)$, where d_i is the size of the ith continuous sector of address space occupied by memory and D is the total size of occupied memory. Two states of occupied memory are regarded as distinct if they are different as vectors.

In modeling inquiries for distinguishing free memory from $z(N) = (n_1, n_2, \ldots, n_r)$ it is assumed that these can enter either simultaneously, i.e. in batches $q(K) =$

$(k_1, k_2, \ldots, k_t)$, or alone, where k_j is the required size of free memory at the $j$the inquiry. We interpret a batch of inquiries also as a vector, i.e.

$$q(K) = (k_1, k_2, \ldots, k_t), \qquad \sum_{j=1}^{t} k_j = K, \qquad k_1 \geq k_2 \geq \cdots \geq k_t,$$

$k_j \in \mathbf{Z}^+$. Two batches of inquiries are assumed to be distinct if they are different as vectors.

The components of the vectors $z(N)$, $g(D)$, $q(K)$ (i.e. n_i, d_m, k_j) are natural numbers. Hence, $z(N)$, $g(D)$, and $q(K)$ may be interpreted as partitions of the numbers N, D, and K, respectively, i.e. $p(N) = (n_1, n_2, \ldots, n_r)$, $p(D) = (d_1, d_2, \ldots, d_l)$, $p(K) = (k_1, k_2, \ldots, k_t)$, where the sectors n_i are defined as the sizes of free sectors of address memory space, the sectors k_j are the required sizes of continuous sectors of free memory in address space or the sizes of memory inquiries, and the sectors d_m are the sizes of continuous sectors of address space occupied by memory. The ranks of the partitions $p(N)$, $p(K)$, and $p(D)$ are defined as, respectively: r, the number of continuous sectors of free memory in address space; t, the amount of inquiries in the queue; and l, the number of continuous sectors of address space occupied by memory.

The interpretation of states of free and occupied memory by unordered number partitions allows us to adequately model external memory fragmentation without taking into account states of its address space, which essentially simplifies the investigation of the process of memory fragmentation. In fact, in our subsequent investigations we will be interested in answering the question: Is there a free continuous sector of address memory space that suffices for fulfilling the entering memory inquiries? In this problem statement we do not need data on states of the address space of free memory.

The representation of batches of inquiries on distinguishing memory by unordered number partitions also does not contradict the practical meaning of the process under investigation. If inquiries on memory goes in batches, then they all must be satisfied simultaneously, and the algorithm of allocating the inquiries can be arbitrary, as can be the order or sequence of distinguishing free memory sectors for them. Hence, the model 5.1 is an adequate representation of both the states of fragmented memory and the system of inquiries on distinguishing memory that can be formed during the process of functioning of the automatic control system.

Using the set of number partitions we can describe the set of all possible states of fragmented free memory of fixed size. As already noted, in the process of functioning of the automatic control system the total size of free computer memory varies between $0 \leq N \leq Q$, where Q is the size of computer memory. Using the following property of the set of number partitions:

$$P(N_1) \cap P(N_2) \cap \cdots \cap P(N_r) = 0,$$

where $P(N_i)$ is the set of partitions of the number N_i, and $N_i \neq N_j \ \forall i \neq j$, we can show that the set of states of fragmented free computer memory of size Q is defined as the set of number partitions

$$Z(Q) = \bigcup_{N=0}^{Q} Z(N) = \bigcup_{N=0}^{Q} \bigcup_{r=1}^{\min(N, Q+1-N)} P_r(N),$$

or (using cardinalities of sets)

$$|Z(Q)| = \sum_{N=0}^{Q} |Z(N)| = \sum_{N=0}^{Q} \sum_{r=1}^{N \wedge (Q+1-N)} |P_r(N)|,$$

where $P(N)$ is the set of number partitions of rank r. The truth of these equalities is confirmed in

THEOREM 5.1. *A partition $p(N) \in P(N)$ corresponds to a state of free computer memory of size Q if and only if $N + r(p) - 1 \leq Q$, where $r(p)$ is the rank of the partition $p(N)$.*

PROOF. Necessity. Let $p_r(N)$ be a partition of a number N of rank r which corresponds to a state of free memory of size N. By the definition of fragmentation, between every two n_i and n_{i+1} there is a segment d_j of occupied memory. Suppose the minimal size of the segments is equal to one ($d_j \geq 1$). Then, clearly, the number of occupied segments will be $l \geq r(p) - 1$. Hence the total size of occupied memory $F \geq r(p) - 1 = r(p-1)$. But since $r(p)$ is the rank of $p(N) \in P(N)$ with $0 \leq N \leq Q$, we have $F \geq Q - N$.

Sufficiency. Let $N + r(p) - 1 \leq Q$. Then the size of occupied memory is $F = Q - N \geq r(p) - 1$. This means that we can find segments of occupied memory which fill all $r(p) - 1$ places between segments of free memory by these $r(p) - 1$ ones. This we need to prove. Then

$$\sum_{N=0}^{Q} |Z(N)| = \sum_{N=0}^{Q} \sum_{p \in P(N)} \chi(p) = \sum_{N=0}^{Q} \sum_{p \in P(N)} \chi((N + r(p) - 1) \leq Q),$$

where χ is the indicator function, taking values as follows:

$$\chi(p) = \begin{cases} 1 & \text{if } p \text{ correponds to a state of free memory,} \\ 0 & \text{if } p \text{ does not correpond to a state of free memory.} \end{cases}$$

It is well known that $|P(N)| = \sum_{r=1}^{N} |P_r(N)|$, hence

$$\sum_{N=0}^{Q} |Z(N)| = \sum_{N=0}^{Q} \sum_{k=1}^{N} \chi((N + r(p) - 1) \leq Q) \sum_{p \in P(N)} 1 =$$

$$= \sum_{N=0}^{Q} \sum_{r=1}^{N} \chi((N + r(p) - 1) \leq Q) |P_r(N)| =$$

$$= \sum_{N=0}^{Q} \sum_{r=1}^{N} \xi(r(p) \leq (Q + 1 - N)) |P_r(N)| = \sum_{N=0}^{Q} \sum_{r=1}^{\min(N, Q+1-N)} |P_r(N)|.$$

Thus, the number of states of free memory of size Q is defined by

$$\sum_{N=0}^{Q} |Z(N)| = \sum_{N=0}^{Q} \sum_{r=1}^{\min(N, Q+1-N)} |P_r(N)|. \qquad \square$$

Studies of the process of memory allocation of automatic control systems also include the process of satisfying inquiries in address space of free computer memory. A formalization of this process is as follows.

MODEL 5.2. A specific feature of memory allocation in computers with segmented organization of programs and data is the indivisibility of entering inquiries on distinguishing memory, i.e. to satisfy each inquiry one needs a continuous sector of address memory space of varying length. This kind of organization of memory allocation is applied in real telecommunication systems of KAMA type, in the allocation of operation memory of multiprocessor computing complexes of ELBRUS type, and in many other systems. Satisfaction of an arbitrary inquiry on memory is here realized by the successive execution of two processes: the process of searching for free continuous sectors of address memory space equal to or exceeding the size of the inquiry, and the process of assigning this free sector to the inquiry.

Both processes may be realized by various algorithms; however, in the course of executing them the found sector of free memory is either completely excluded from the list of free sectors (if the sizes of the sector and the inquiry coincide), or in the list of free memory the remainder of free memory, the difference between the sizes of the sector and the inquiry, is taken into account (if a free sector of memory of size larger than that of the inquiry is chosen). As a consequence we can show that more than one inquiry on memory can be satisfied by a single sector of free memory. This means that processes of satisfying inquiries on memory can be modeled by the notion of packability of number partitions. In fact, suppose that, in correspondence with the model 5.1, the inquiries, interpreted as partitions $(k_1, \ldots, k_t) \vdash k$, must necessarily be satisfied in memory, for which the free sectors correspond to parts of the partition $(n_1, \ldots, n_r) \vdash n$ and $k \leq n$.

In correspondence with the definition of packability, the partition $(k_1, \ldots, k_t)$ can be packed into the partition $(n_1, \ldots, n_r)$ if a part k_i of $(k_1, \ldots, k_t)$ can be grouped into r batches (each part k_i goes into a single batch, and empty batches are discarded) such that after addition of all parts k_i in each batch we obtain r numbers $p_i \leq n_i$ $i = 1, \ldots, r$. Moreover, in a concrete packing process each n_j in $(n_1, \ldots, n_r)$ is used at most once, i.e. the fragment of size n_j, in which a batch of inquiries occupies size $p_j \leq n_j$, is not used anymore for distributing inquiries k_i, even if $n_j - p_j > 0$. Consequently, the notion of packability of partitions is an adequate interpretation of the process of satisfying inquiries on free computer memory.

As an illustration we consider a concrete numerical model.

EXAMPLE 5.1. Suppose the batch of inquiries on memory consists of sizes $(5, 2, 1)$, while the system of sectors of free memory consists of sizes $(6, 3, 3)$. Then the inquiries can be simultaneously satisfied, and, moreover, in a nonunique manner: (6 contains 5 and 1, 3 contains 2, 3 contains 0), (6 contains 5, 3 contains 2 and 1, 3 contains 0), (6 contains 5, 3 contains 2, 3 contains 1). This manner of grouping inquiries for distributing them in sectors of free memory precisely reflects the actual work of algorithms of dynamic memory allocation.

Using this terminology, we consider the application of results of solving extremal combinatorial problems in the design of methods for controlling computer memory allocation.

2. Design of algorithms for controlling computer memory allocation

There are various algorithms used to stave off the influence of external fragmentation of computer memory. However, in the realization of any such algorithm, in order to identify failure in satisfying inquiries one consults the whole list of free memory sectors. Moreover, this list is consulted for each inquiry separately. This consultation of the list of free memory involves an expense of computing resources of the central processor (processor time). If as the result of the consultation a necessary part of free memory is not found, the entering inquiry cannot be satisfied in this case, and the processor time used in the consultation turns out to have been used in vain. This algorithm is used in practice in all Soviet, and in the majority of foreign, computers. The great intensity of the flow of inquiries on distinguishing memory in the process of functioning of a computer, and also the high frequency of the memory situations considered above, reduce the productivity of computing systems in the large. For example, in correspondence with estimates in [90], the loss incurred by control of memory makes up around 12% of the overall productivity of a system working in a time-sharing regime.

For computing systems in whose process of functioning there does not arise a queue of inquiries on distinguishing free memory, one may exclude the nonuseful expense of processing time for consulting the list of free memory by comparing the sizes of entering inquiries with the size of the maximal sector of free memory (this quantity must be stored in the system and must be dynamically corrected in the process of functioning).

However, in the functioning of simultaneous multiprocessor computing complexes queues on distinguishing memory do arise. They are formed because of conflicts when addressing common data (the list of free memory, etc.), and also in the realization of mechanisms of re-allocation and reorganization of memory [48]. Such queues (batches of inquiries) can serve as a source of a priori information on the nature of the flow of inquiries on distinguishing memory. This makes it possible to increase the efficiency of the use of computer memory on account of the possibility of a more rational planning of distributing information objects in address space of free memory, and allows a reduction of expense in processor time for controlling memory allocation on account of a reduction of the number of consults of the list of free memory. Nevertheless, in the design of modern algorithms for memory allocation, queues of inquiries are not taken into account.

A partial solution of the problem of designing algorithms for memory allocation, taking into account possible formation of queues of inquiries, is made possible by the results of solving extremal combinatorial problems on packability of number partitions (Theorem 2.1). In terms of the models 5.1 and 5.2, the main problem in satisfying batches of inquiries consists in establishing the possibility of packing the partition $(k_1, \ldots, k_t)$, interpreted as the sizes of the inquiries in the batch, into the partition $(n_1, \ldots, n_r)$, interpreted as the sizes of fragments of free memory. According to Theorem 2.1, the partition $(k_1, \ldots, k_t)$ can be packed in the partition $(n_1, \ldots, n_r)$ if

$$t \geq \max \left(k - \left] \tfrac{n}{r} \right[+ 1, 1 \right),$$

where t is the number of inquiries in the batch and $k = \sum_{i=1}^{t} k_i$, $n = \sum_{j=1}^{r} n_j$.

In the formal statement of this problem, finding the quantity $\max(k -]n/r[+ 1, 1)$ means that when determining the possibility of satisfying each inquiry of a batch with sizes $(k_1, \ldots, k_t)$ in fragmented address space of free memory it is not required to consult t times the list of free memory. For this it suffices to preserve in the computer only data regarding the total size and amount of parts of free memory, and also data regarding the amount of inquiries in a batch and their total size. Precisely the use of these data in determining the possibility of satisfying entering batches of inquiries ensures a complete absence of expense in processor time for nonuseful consultation of the list of free memory, since if the packability conditions are fulfilled, then the result of consulting the list of free memory always gives free sectors of address memory space which can satisfy each inquiry in the queue.

However, determining $\max(k -]n/r[+ 1, 1)$ is not the final answer to the problem posed. Every new result related to the solution of extremal combinatorial problems on the set of number partitions, partially ordered by packability, always gives also a solution of the problem of designing an algorithm realizing this packing. Therefore it is now necessary to determine or construct an algorithm that, when the packability conditions hold, would guarantee a complete allocation of the inquiries in the batch in computer memory (would guarantee packability of the parts of the partitions). By definition, a packing of a part k_i of a partition $(k_1, \ldots, k_t)$ into a partition $(n_1, \ldots, n_r)$ transforms the partitions to the forms $(k_1, \ldots, k_{i-1}, k_{i+1}, \ldots, k_t)$ and $(n_1, \ldots, n_j - k_i, \ldots, n_r)$.

This means that after packing each part k_i, the rank of $(k_1, \ldots, k_t)$ is less by one. For $n_j = k_i$ the rank r of $(n_1, \ldots, n_r)$ is also less by one, while if $n_j > k_i$ it is unchanged. This interpretation of packability of parts of partitions adequately formalizes the working of an algorithm for memory allocation. In the proof of the following proposition, allowing us to choose an algorithm looked for, we understand the procedure of packing in just this manner.

PROPOSITION 5.1. *If partitions $p_r(n) = (n_1, \ldots, n_r)$ and $q_t(k) = (k_1, \ldots, k_t)$ satisfy the condition $t \geq \max(k -]n/r[+ 1, 1)$, then the packing $(k_1, \ldots, k_t) \subset (n_1, \ldots, n_r)$ is guaranteed by every algorithm that distributes parts equal to one at the last stage.*

PROOF. It suffices to consider the case $n = k$. Let $k_i < n_j$. Clearly, after packing k_i in n_j the rank r does not change, and complete packability is guaranteed if $k -]n/r[\geq k - k_i -](n - k_i)/r[+ 1$, which, in turn, is equivalent to $](k - (r - 1)k_i)/r[\geq](k + r)/r[$. The last inequality clearly holds if $k \geq]r/(r - 1)[= 2$. If $k_i = n_j$, the sought inequality has the form $k -]k/r[\geq k - k_i -](k - k_i)/(r - 1)[+ 1$. $\square$

We assume that $](k + (r - 1)k_i)/(r - 1)[\geq]k/r[$, but that this inequality does not hold if $k_i \geq 2$. Thus, the proposition has been proved.

Consequently, if the condition of Theorem 2.1 (the condition for packability of partitions) holds, then by Proposition 5.1, to satisfy inquiries in a group we need not only require them to be ordered, but also require ordering by magnitude of the sizes of the sectors of free memory in the list. In this case it is only necessary to allocate the inquiries $k_i = 1$ at the last stage. A rule for choosing an n_j to satisfy the inquiry of size k_i can be given as follows:

$$j = \min\{j : k_i \leq n_j\}, \qquad 1 \leq j \leq r,$$

where r is the amount of sectors of free memory, which are, possibly, not ordered by magnitude. This rule is realized by the `first-fit` algorithm for allocating memory [41], which is the fastest allocation algorithm, i.e. requires for its working a minimum of expense in central processor time. This means that by first verifying the fulfillment of the packability conditions before starting the algorithm allocating the memory, we can without consulting the list of free memory determine the possibility of satisfying an entering batch of inquiries. If these conditions are fulfilled, then the `first-fit` algorithm (with inessential reworking, in accordance with Proposition 5.1) will completely satisfy the inquiries in the batch, without the use of some or other means of reorganizing memory.

We must note that fulfillment of the packability conditions for a batch of t inquiries with total size k indicates the presence of a reserve of free memory of volume at least $r(k-t)-k+r$, where r is the amount of sectors of free memory. Clearly, in computing systems that characteristically process inquiries on distinguishing memory that are large in size (tens or more quanta), the working according to such an algorithm may lead to the appearance of a large reserve of free memory. However, for systems in which the inquiries on memory are not large and are of the same quantum, in the process of their functioning a reserve of free memory will be created whose size will dynamically reduce in dependence of the nature of the flow of inquiries on distinguishing memory (the more entering inquiries, the larger the size of the reserve of free memory). This reserve may be used by the computer for protection from deadlock situations in operation memory. The quantity of this reserve may be reduced by using the principle of complete packability (see Chapter 2) to verify the packability conditions. In this case we need additional information on the sizes of the inquires in batches, although the solution of this problem is polynomially complex.

The given application of results of solving extremal combinatorial problems is not the only one. It and other extremal results may be used for studying the process of job execution in automatic control systems, in choosing the sizes of operation and external computer memory, and in the analysis of specifics of software structures of automatic control systems.

3. A combinatorial model for studying the process of job execution in an automatic control system

The functioning of an automatic control system is made up from a set of random processes and phenomena of various complexities. Their study is basic for the increase of the efficiency of organization of the computing process of an automatic control system in the large. However, many of these processes and phenomena can only with great difficulty be modeled by analytical methods of study. As a result the analytical models developed turn out to be unsuitable for even obtaining bounds on the values of the parameters under investigation.

The use of methods from combinatorial analysis for studying the process of functioning of an automatic control system makes it possible to built from combinatorial schemes more adequate formal models of the elements, processes, and phenomena under investigation. The unification of these models into combinatorial schemes on the basis of common parameters guarantees, on the level of estimation of parameter values, the possibility of an analysis of the mutual influence of parameters, as well as of their influence on the process of functioning of the automatic control system. To substantiate this thesis, we consider a combinatorial model for job execution in an automatic control system.

MODEL 5.3. A *job* in an automatic control system is a function to be realized by it by executing a program or sequence of programs. If some function in an automatic control system is realized, in dependence on initial information, by various sequences of programs, then in the proposed combinatorial model such a distinction in realization is regarded as a distinction of functions to be realized, i.e. in the model it is assumed that each function to be realized by the automatic control system is executed by a rigidly fixed sequence of programs. This assumption does not impose any restriction on the generality of using the model, since in designing the automatic control system we always have the possibility of such a detailed representation of the functions to be realized by it.

The proposed model allows us to investigate the automatic control system under the following restrictions on the process of using it:

- the job serving mechanism in the automatic control system is such that for each terminal a subsequent job can be initialized only if the previous job given with the same terminal is completely executed;
- the software structure of the automatic control system is fixed and intended for realizing finitely many functions processing information, and is given when designing the system.

This organization of the functioning of the automatic control system is typical in a number of systems of a similar class, which also ensures the generality of the proposed model.

Suppose the automatic control system is intended for serving f terminals and realizing various functions, by executable jobs $z \in Z$, where Z is the set of all functions realisable by the automatic control system. Let Q_i be a set of jobs, each of which can be initiated at the terminal $i \in [f] = \{1, \ldots, f\}$, where any Q_i and Q_j may intersect. We call Q_i a *description of the functional assignment* of terminal i. Taking in correspondence with the restrictions of the mechanism of job execution in the automatic control system, it allows us to assume that at any time instant the automatic control system can execute at most f jobs simultaneously. Then the set of distinct families of jobs that can be simultaneously executed in the automatic control system is defined as a direct product $Q = \prod_{i=1}^{f} Q_i$ whose elements $q = (z_1, \ldots, z_f) \in Q$ are called *complete families of jobs* for the designed allocation of functions between terminals in the automatic control system and the given sets Q_i (z_i are jobs intended to be induced by terminal i).

In the process of functioning of the automatic control system the formation of any $q \in Q$ is preceded by a set of different successive states of the automatic control system, characterized by the simultaneous execution of jobs. Let $q = (z_1, \ldots, z_f)$; for this q we consider two states of the automatic control system, characterized by the execution of respectively one (z_1) or two simultaneous (z_1, z_2) job(s). It is clear that in the process of functioning of the automatic control system the sequences of its states transforming the system from (z_1) to (z_1, z_2) can be very distinct. Here we are not interested in the number of times that the system is in a certain state, since our problem is the investigation of the set of all possible states of the automatic control system. To conduct this investigation it suffices to fix only the possibility that the automatic control system is in some state. This greatly simplifies the model for describing the set of all states of the automatic control system, taking (z_1) and (z_1, z_2) to be adjacent states if there is a transition of the automatic control system from (z_1) to (z_1, z_2), and using the following approach to describe the whole set of states of the system.

Let $Q(q) = \{q_i\colon q_i \subset q\}$, $q_i = (z_{i_1}, \ldots, z_{i_l})$, $(i_1, \ldots, i_l) \in 2^{[f]}$, where $2^{[f]}$ is the Boolean of the set $[f]$. In correspondence with the definition of Boolean (see Chapter 1), it is clear that the elements of $Q(q)$ are all possible pairings from $q = (z_1, \ldots, z_f)$. Hence $Q(q)$ can be interpreted as the set of all states of the automatic control system, characterized by the execution of one, two simultaneous, three simultaneous, etc. up to f jobs from $q \in Q$. However, using $Q(q)$ we can also interpret all possible queues of jobs when forming a concrete full family $q \subset Q$. In fact, let $|q_i|$ be the amount of jobs making up $q_i \in Q(q)$. To form $q \in Q$, in the system there must already been realized $f - |q_i|$ such jobs, which is the complement $\bar{q}_i$ of q_i in q, i.e. $\bar{q}_i = q \setminus q_i$. Using the rule for constructing $Q(q)$, it is not difficult to prove that if $\bar{q}_i = q \setminus q_i$ and $q_i \in Q(q)$, then $\bar{q}_i \in Q(q)$. So, by giving the bijective map $\phi\colon Q(q) \to Q(q)$ such that $\phi(q_i) = \bar{q}_i$, we obtain pairs of elements $(q_i, \bar{q}_i)$, characterizing each state of the automatic control system by simultaneous execution of the jobs in q_i and the corresponding state of the queue of jobs to be executed when the system transfers from q_i to the state characterized by the jobs from $z \in Q$. Using this approach to define intermediate states for all $q \in Q$, we obtain a set $W(Q) = \{(w, \bar{w})\colon \bar{w} = q \setminus w, w \in Q(q), q \in Q\}$., whose elements allow us to obtain a priori information for estimating the values of certain parameters of the functioning of the automatic control system.

The cardinality of the set $W(Q)$, or the amount of elements in it, determines the amount of calculations necessary for studying the whole variety of states of the automatic control system. Using the rule for constructing complete families $q \in Q$ and the rule for constructing $Q(q)$, it is not difficult to prove that the cardinality of the set $W(Q)$ is defined by $|W(Q)| = 2^f \cdot \prod_{i=1}^{f} |Q_i|$, where 2^f is the cardinality of the Boolean of the set $[f] = \{1, \ldots, f\}$ and $|Q_i|$ is the cardinality of Q_i. There are various applications of this model in studies of the functioning of automatic control systems. As an example we consider its use in estimating from above the necessary size of operation computer memory. However, for this we need consider another series of combinatorial models, formalizing the process of memory allocation and the influence of external fragmentation.

4. Combinatorial models for estimating the necessary size of computer memory

The size of the operation computer memory has an essential influence on the capacity of the automatic control system. If the size of operation memory is small, then in the process of functioning of the system part of the central processor time is lost on controlling memory allocation. Increase in operation memory increases the productivity of the system without any change in the programmatic treatment of data. Memory will always be a key factor in the productivity of computers. J. von

Neumann established this in his 1946 memoir, and it is true today [92].

In the design of real-time automatic control systems problems of estimating the size of operation memory are given attention at practically all levels of creation of the system. To this end complex imitation models have been created with whose aid one studies, basically, the behavior of the system under peak loading, i.e. in time periods when the average quantity of flow of jobs in the system takes a maximal value [14], [92]. These studies require an essential amount of computing resources in the development of the system, and increase the time for creating it. However, such studies are necessary, since precisely under peak loadings a real-time automatic control system must remain running.

We must also note that efficient functioning of a real-time automatic control system is not possible without the fulfillment of a very important condition: as a design outcome, software should 'correspond' to the machine; it should be designed such that it does not reduce the productivity of the machine and of the system in the large [26], [92]. The approaches to solving the problem of controlling memory allocation used nowadays are such that a final solution emerges, as a rule, only at the stage of running the system.

The above-given short analysis of requirements on methods for controlling the allocation of operation computer memory in real-time automatic control systems allows us to draw the following conclusions:

- a method solving the problem of designing control of operation memory allocation must minimize the expense of computing resources, and must guarantee that a theoretically substantiated algorithm can be obtained at the stage of technically designing the automatic control system;
- a method for controlling memory allocation must be designed while taking into account the specifics of the software structure of the system;
- a method for controlling the operation computer memory must minimize the expense in processing time for allocation, and must guarantee efficient execution of functions of the real-time automatic control system under peak loadings on the system;
- a method for controlling the operation computer memory must impose fewest possible restrictions in satisfying inquiries on distinguishing memory, and must guarantee fast emptying of segments of memory not taking part in the computing process.

For the solution of these problems we propose a number of combinatorial models. We have chosen the sequence of presenting these models in correspondence with the increase in a priori information on the process of functioning of the operation computer memory of the automatic control system. In our models we will also consider the possibility of batch satisfaction of inquiries, i.e. when the inquiries on distinguishing free memory enter in batches. The advantage of a batch method for satisfying inquiries lies in the presence of additional a priori information on the nature of the flow of inquiries in memory, which is here taken into account by way of considering batches of inquiries ordered by size. We also consider a model of the process of allocating operation computer memory when a solitary method of satisfying inquiries on distinguishing memory is realized, i.e. when inquiries are satisfied in the order of their appearance.

MODEL 5.4. We consider the functioning of a computer in which inquiries on distinguishing memory occur in batches. Let the sizes of the inquiries in a batch entering at an arbitrary moment of time correspond to parts of a partition $(k_1, \ldots, k_t) \vdash k$ At this time moment the free memory is represented by r sectors, with total size n. Then by the principle of complete allocation (see Chapter 2) the computation of the quantities

$$n(k_1, \ldots, k_t; r) = \max_{1 \leq i \leq t} \left(\sum_{j=1}^{i} k_j + (k_i - 1)(r - 1) \right),$$

under the condition $k_1 \geq \cdots \geq k_t$, allows us to find a total amount of free memory which, being represented by any partition on the r continuous free sectors in address space, makes it possible to completely distribute all $k_1, \ldots, k_t$ in it without reallocation and reorganization. The formula of the principle of complete allocation implies that for solving this problem we do not need information on the sizes of the free sectors of memory, and thus for none of the inquiries in the batch need we examine the list of free computer memory. To this end it suffices to verify the validity of the inequality $n \geq n(k_1, \ldots, k_t; r)$, where r is the amount of fragments by which the free memory of size n is represented.

If the inequality is satisfied, then the proof of the principle of complete allocation implies that for satisfying the inquiries $(k_1, \ldots, k_t)$ we can use any algorithm of dynamic memory allocation that uses the fact that the inquiries are ordered by decreasing sizes. In other words, all inquiries of the batch $k_1, \ldots, k_t)$ can be simultaneously satisfied by the free memory $(n_1, \ldots, n_r)$, e.g. using the first-fit algorithm, if the inquiries are chosen from the queue in the order of decreasing sizes. Hence, the principle of complete allocation may be used in the design of methods for dynamic computer memory allocation in automatic control systems.

EXAMPLE 5.2. Suppose the sizes of the inquiries are represented by parts of the partition $(k_1, \ldots, k_t) = (22, 13, 12, 8, 4, 2, 2, 1) \vdash 64$, while the continuous sectors of address space in free memory correspond to $(n_1, \ldots, n_r) = (23, 21, 21, 20) \vdash 85$. Then

$$n(k_1, \ldots, k_t; r) = n(22, 13, 12, 8, 4, 2, 2, 1; 4) = 85,$$

and hence we can place all represented inquires, e.g. as follows: $(23 = 22 + 1, 21 = 13 + 8, 21, 20 = 12 + 4 + 2 + 2)$. If, however, $(n_1, \ldots, n_r) = (22, 21, 21, 20) \vdash 84$, then the principle of complete allocation does not yet imply the required packing, despite the fact that packing is possible. This characterizes 'zones of indeterminacy' in extremal combinatorial bounds.

MODEL 5.5. Suppose we have m batches of inquiries on distinguishing memory. The sizes of the inquiries in the jth batch correspond to parts of the partition $(k_1^{(j)}, \ldots, k_{t_j}^{(j)})$ $(j = 1, \ldots, m)$. In agreement with the principle of complete allocation it is not difficult to prove that computing the quantity

$$\max_{1 \leq j \leq m} n(k_1^{(j)}, \ldots, k_{t_j}^{(j)}; r),$$

under the conditions $k_1^{(j)} \geq \cdots \geq k_{t_j}^{(j)}$, makes it possible to find the total size of free memory every partition of which in at most r continuous address sectors guarantees that all inquiries on memory in the jth batch can be satisfied $(1 \leq j \leq m)$. The algorithm for placing the inquiries is the same as in the model 5.4.

MODEL 5.6. Suppose the set of possible states of occupied computer memory is interpreted as the set of number partitions $(d_1^{(i)}, \ldots, d_{v_i}^{(i)})$, $i = 1, \ldots, l$, where every part of the ith partition $d_r^{(i)}$ corresponds to the size of the rth occupied sector of address memory space, and v_i is the amount of occupied sectors, represented by the parts of the ith number partition. For each ith state of occupied memory there is known a set of batches of inquiries, in the course of functioning of the automatic control system each of which requires simultaneous satisfaction of all inquiries in it for the ith state of occupied computer memory.

Suppose the set of number partitions $(k_1^{(j)}, \ldots, k_{t_j}^{(j)})^{(i)}$ corresponds to this set of batches of inquiries for the ith state of occupied memory; the parts of these partitions correspond to the sizes of the inquiries on memory, and the rank t_j is the amount of inquiries in the jth batch $(i = 1, \ldots, l; j = 1, \ldots, m_i)$. Taking into account the influence of external fragmentation, we obtain that in the process of functioning of the automatic control system when satisfying each batch of inquiries corresponding to the ith state of occupied memory, the free computer memory turns out to be 'scattered' across at most $v_i + 1$ continuous sectors of address space, where v_i is the number of occupied sectors of memory corresponding to the ith state. Then an upper bound for the necessary size (V) of computer memory that, in the process of functioning of the automatic control system, would suffice for satisfying any entering batch of inquiries, taking into account the principle of complete allocation, can be computed by the following expression:

$$V = \max_{1 \leq i \leq l} \left(\max_{1 \leq j \leq m_i} n((k_1^{(j)}, \ldots, k_{t_j}^{(j)})^{(i)}; v_i + 1) + \sum_{r=1}^{v} d_r^{(i)} \right),$$

for $k_1^{(j)} \geq \cdots \geq k_{t_j}^{(j)}$, where $n((k_1^{(j)}, \ldots, k_{t_j}^{(j)})^{(i)}; v_i + 1)$ is defined by the formula for complete allocation for each partition $(k_1^{(j)}, \ldots, k_{t_j}^{(j)})^{(i)}$, and $\sum_{r=1}^{v} d_r^{(i)}$ is the total size of occupied memory under satisfaction of the inquiries in a batch.

This approach to choosing the size of operation computer memory of an automatic control system, as in the preceding models, assumes that it is possible to use algorithms of dynamic memory allocation which take into account the batches of inquiries and ensure satisfaction of the inquiries in these batches in the order of decrease of their sizes. However, nowadays the majority of computers in automatic control systems do not use the batch method of satisfying inquiries on memory. The construction of an analogous model for estimating the required size of operation computer memory under realization of an individual method of satisfying inquiries is made possible by a somewhat different formal combinatorial model of entering batches of inquiries in the process of functioning of the automatic control system. Suppose we have a batch $(k_1, \ldots, k_t)$ of inquiries on distinguishing memory, which must be satisfied in the order of entrance. In a finite time span, all inquiries must be simultaneously satisfied in free memory, which at the moment the first inquiry of the batch enters is represented by r continuous sectors of address space. Clearly, the individual satisfaction of the inquiries in the batch under consideration presupposes that the numbers $k_1, \ldots, k_t$ are in arbitrary order. In accordance with Lemma 2.4, if

$$f(k_1, \ldots, k_t; r) = \max_{1 \le i \le t} \left(\sum_{j=1}^{i} k_j + (k_i - 1)(r - 1) \right), \qquad k_i \in \mathbf{N},$$

then for any sequential ordering of the numbers $k_1, \ldots, k_t$ we have the inequality

$$f(k_1, \ldots, k_t; r) \ge n(k_1, \ldots, k_t; r).$$

It is then clear that to estimate the required size of free computer memory under individual satisfaction of inquiries it is necessary to find the maximum of $f(k_1, \ldots, k_t; r)$ on the whole set of permutations of the numbers $(k_1, \ldots, k_t)$. By Lemma 2.4 this maximum is equal to

$$k + \left(\max_{1 \le i \le t} k_i - 1 \right)(r - 1), \qquad \text{where } k = \sum_{i=1}^{t} k_i.$$

However, this result is not the final solution to the problem. Before turning to the description of a model for studying the individual method of satisfaction of inquiries, we must formulate the algorithm by which we can realize the packing. The algorithm realizing packability for an arbitrary ordering of the parts of a partition $(k_1, \ldots, k_t)$ is as follows.

PROPOSITION 5.2. *Suppose we are given partitions* $(k_1, \ldots, k_t) \vdash k$ *and* $(n_1, \ldots, n_r) \vdash n$ *for which the inequality* $n \ge k + (\max_{1 \le i \le t} k_i - 1)(r - 1)$ *holds. Then* $(k_1, \ldots, k_t) \subseteq (n_1, \ldots, n_r)$, *and this packing is realized by the following algorithm:*

- *the parts k_i for distribution are chosen in the order of their labeling;*
- *each part k_j is distributed in the first part n_i that has appropriate size, i.e. n_i is chosen by the rule*

$$i = \min(i: \ k_j \le n_i), \qquad 1 \le i \le r, \qquad 1 \le j \le t;$$

- *when choosing an $n_i > k_j$, there remains a single part $n_i - k_j > 0$.*

PROOF. We proceed by induction with respect to t. For $t = 1$ the main inequality takes the form $n \ge r k_1 - r + 1$, and the required is fulfilled by Dirichlet's principle. We make the induction step. Assume that the required is fulfilled up to $t - 1$ inclusive; we show that it is also true for t. The packing $(k_1, \ldots, k_t) \subseteq (n_1, \ldots, n_t)$ follows from the principle of complete allocation and the fact that $n(k_1, \ldots, k_t; r) \le k + (\max_{1 \le i \le t} k_i - 1)(r - 1)$ by Lemma 2.4. So, for k_1 we can always find an $n_i \ge k_1$. After packing the part k_1, the required will follow from fulfillment of the packability condition for the partitions $(k_1, \ldots, k_t)$ and $(n_1, \ldots, n_i - k_1, \ldots, n_r)$. To prove this packability we use the induction hypothesis. To this end it suffices to verify the truth of the inequality

$$n - k_1 \ge k - k_1 + \left(\max_{2 \le i \le t} k_i - 1 \right)(v - 1),$$

where

$$v = \begin{cases} r & \text{for } n_i > k_1, \\ r - 1 & \text{for } n_i = k_1. \end{cases}$$

In fact, by requirement,

$$n \geq k + \left(\max_{1 \leq i \leq t} k_i - 1\right)(r-1) + \left(\max_{1 \leq i \leq t} k_i - 1\right)(v-1),$$

and thus

$$n - k_1 \geq k - k_1 + \left(\max_{2 \leq i \leq t} k_i - 1\right)(v-1),$$

as required. $\square$

MODEL 5.7. Let the set of possible states of occupied computer memory be represented by the set of number partitions $(d_1^{(i)}, \ldots, d_{v_i}^{(i)}) \vdash d^{(i)}$, $i = 1, \ldots, l$. For each element of this set there are m_i batches of inquiries on distinguishing memory known, represented by corresponding partitions $(k_1^{(j)}, \ldots, k_{t_j}^{(j)})^{(i)}$, $j = 1, \ldots, m$, $k_r^{(i)}$ being the size of the rth inquiry in the jth batch, and t_j being the amount of inquiries in the jth batch. In the process of functioning of the computer of the automatic control system the inquiries of any jth batch enter in arbitrary order, and are individually satisfied in correspondence with the order of entrance and require simultaneous distribution in memory. At the moment of entrance of the first inquiry of each jth batch, corresponding to the ith state of occupied memory, the free computer memory is represented by $v_i + 1$ continuous sectors of address space. Then by Lemma 2.4 this model allows us to give an upper bound on the required size (V') of operation computer memory,

$$V' = \max_{1 \leq i \leq l}\left(\max_{1 \leq j \leq m_i}\left(k^{(i,j)} + \left(\max_{1 \leq v \leq t_j^{(i)}} k_v^{(i,j)} - 1\right)v_i\right) + \sum_{r=1}^{v} d_r^{(i)}\right),$$

where $k_v^{(i,j)}$ is the size of the vth inquiry on memory in the jth batch, corresponding to the ith state of occupied memory; $t_j^{(i)}$ is the amount of inquiries in the jth batch, corresponding to the ith state of occupied memory;

$$k^{(i,j)} = \sum_{v=1}^{t_j^{(i)}} k_v^{(i,j)}.$$

Proposition 5.2 makes it possible to prove that when the computer of the automatic control system functions with operation memory of size V', then satisfaction of any inquiry is guaranteed by the first-fit algorithm of dynamic memory allocation.

We must note that for the computation of V and V' in this case we do not need information on the sizes of the continuous sectors of free address space. This is in agreement with the formal interpretation of the principle of complete allocation. However, if at some stage of the design of the automatic control system additional a priori information on the functioning of the system element under investigation becomes available, then, using other results concerning the packability of number partitions, we can refine the values of the parameters under investigation. Suppose that the sizes of the inquiries in a batch entering at an arbitrary moment of time correspond to the parts of a partition $(k_1, \ldots, k_t) \vdash k$. At this moment of time the free memory is represented by r sectors with total size n, and, in distinction from the models 5.4–5.7, the sizes of all r sectors of free memory are known, which gives corresponding parts of a partition $(n_1, \ldots, n_r) \vdash n$. This information may be used to improve the extremal bounds of the required size of operation computer memory guaranteeing satisfaction of the inquiries on memory in batch service. In fact, if

$$n_1 \geq \cdots \geq n_r,$$

$m = \min(m: \sum_{i=1}^{m} n_i \geq k)$, and if the inequality

$$\max_{m \leq l \leq r} \left(\sum_{i=1}^{l} n_i - n(k_1, \ldots, k_t; l) \right) \geq 0$$

is satisfied, then all inquiries on memory $(k_1, \ldots, k_t) \vdash k$ can be simultaneously satisfied in the fragmented address space of free memory with sizes of free sectors to which correspond the parts of the partition $(n_1, \ldots, n_r) \vdash n$. To clarify this we consider an example.

EXAMPLE 5.3. Let $t = r = 3$, $(k_1, \ldots, k_t) = (5, 2, 1) \vdash 8$, $(n_1, \ldots, n_r) = (6, 3, 3) \vdash$ 12. Then $n(5, 2, 1; 3) = 13 > 12 = 6 + 3 + 3$, and thus a straightforward use of the principle of complete allocation, without taking into account information on the sizes of the sectors of free memory, will not guarantee satisfaction of the inquiries. However, for $l = 2$ we find that $n(5, 2, 1; 2) = 9 = 6 + 3$, and thus, by the principle of complete allocation, $(5, 2, 1) \subseteq (6, 3)$. Since

$$(n_1, \ldots, n_l) \subseteq (n_1, \ldots, n_l, \ldots, n_r),$$

we have $(5, 2, 1) \subseteq (6, 3) \subseteq (6, 3, 3)$. Hence, by transitivity of packability we find $(5, 2, 1) \subseteq (6, 3, 3)$, so that after all satisfaction of the inquiries is guaranteed.

On the other hand, if in the clarification of the possibility of satisfying entering batches of inquiries on memory we take into account the sizes of the continuous free sectors of address space, then, using the extremal result from Theorem 2.3, we can also refine the upper bound on the free memory which is sufficient to guarantee satisfaction of the entering batches of inquiries. We demonstrate the use of Theorem 2.3 by constructing a number of models.

MODEL 5.8. Consider a computer in an automatic control system in which inquiries on distinguishing free memory enter in batches. To determine the possibility of satisfying all inquiries in a batch in fragmented address space of free memory we may use Theorem 2.3, whose essence lies in finding the quantities

$$n(k_1, \ldots, k_t; n_2, \ldots, n_r) = \max_{1 \leq i \leq t} \left(\sum_{j=1}^{i} k_j + \sum_{l=2}^{r} \min(n_l, k_i - 1) \right),$$

$r \leq \sum n_j$, under the conditions $k_1 \geq \cdots \geq k_t$. If $n(k_1, \ldots, k_t; n_2, \ldots, n_r) \leq \sum_{l=2}^{r} n_l$, then $(k_1, \ldots, k_t) \subseteq (n_1, \ldots, n_r)$, where $(k_1, \ldots, k_t) \vdash k$, $(n_1, \ldots, n_r) \vdash n$ are partitions of k and n, respectively.

The formal source of this result becomes apparent when interpreting the number partitions $(k_1, \ldots, k_t)$, $(n_1, \ldots, n_r)$, and also packability of number partitions, in accordance with the definitions in models 5.1 and 5.2. To determine the possibility of simultaneous satisfaction of inquiries in a batch, whose sizes are interpreted as parts of a partition $(k_1, \ldots, k_t) \vdash k$, in fragmented address space of free memory, interpreted by the partition $(n_1, \ldots, n_r) \vdash n$, we have to verify the truth of the inequality $n(k_1, \ldots, k_t; n_2, \ldots, n_r) \leq n$. Here the quantity $n(k_1, \ldots, k_t, n_2, \ldots, n_r)$ is the minimally necessary total size of free memory which, being represented by r continuous sectors of address space with sizes $p_1, \ldots, p_r$, respectively, with $p_i \leq n_i$, $i = 2, \ldots, r$, guarantees the possibility of simultaneous satisfaction of the inquiries of sizes $k_1, \ldots, k_t$. Hence, if for an incoming batch of inquiries on memory $(k_1, \ldots, k_t)$ the inequality

$$n(k_1, \ldots, k_t; n_2, \ldots, n_r) \leq \sum_{j=1}^{r} n_j$$

holds, then all these inquiries can be simultaneously satisfied in fragmented address space of free memory $(n_1, \ldots, n_r)$ without application of the means of reorganization or re-allocation. As in the models 5.4–5.7, for determining the possibility of satisfying a batch of t inquiries we need not consult the list of free memory t times. Use of information regarding the sizes of continuous sectors of address space of free memory requires in total one consultation of the list of free memory in order to determine whether t inquiries can be simultaneously satisfied. If the inequality $n(k_1, \ldots, k_t; n_2, \ldots, n_r) \leq \sum_{j=1}^{r} n_j$ holds, the guaranteed allocation of each inquiry k_i is ensured by any imaginable algorithm for dynamic memory allocation that uses in its realization the fact that the inquiries are ordered by decreasing size, i.e. $k_1 \geq \cdots \geq k_t$. For example, all inquiries in a batch $(k_1, \ldots, k_t)$ can be satisfied in fragmented address space of free memory with sizes of sectors $(n_1, \ldots, n_r)$ using the first-fit algorithm, if the inequality $n(k_1, \ldots, k_t; n_2, \ldots, n_r) \leq \sum_{j=1}^{r} n_j$ holds and the order of choosing the inquiries from the queue corresponds to $k_1 \geq \cdots \geq k_t$. In other words, if we use Theorem 2.3 as criterion to determine the possibility of simultaneous satisfaction of inquiries in the control of computer memory allocation, then, independently of the rule of choosing the free memory sectors of sizes n_j over which the inquiries k_i are to be distributed, all inquiries will be distributed if they are chosen from their queue in the order of decreasing size. Hence, Theorem 2.3 can be used also to design methods of dynamic computer memory allocation in an automatic control system. As a clarification we consider the same example as in model 5.2. In it $n(5, 2, 1; 3, 3) = 11 < 12 = 6 + 3 + 3$, and hence the required distribution can be realized. Moreover, from the determination of an extremal bound for $n(k_1, \ldots, k_t; n_2, \ldots, n_r)$ it follows that not only a packing

$$(k_1, \ldots, k_t) \subseteq (n_1, \ldots, n_r) \vdash n = n(k_1, \ldots, k_t; n_2, \ldots, n_r)$$

holds, but also the packing of $(k_1, \ldots, k_t)$ into any partition $(p_1, \ldots, p_r) \vdash n$ for which $p_i \leq n_i$, $i = 2, \ldots, r$. So, in the already considered case $(k_1, \ldots, k_t) = (5, 2, 1)$ and $(n_1, \ldots, n_r) = (6, 3, 3)$ we have $n(5, 2, 1; 3, 3) = 11$, and hence the partition $(5, 2, 1)$ can be packed into the following rank-3 partitions of the number 11: $(9, 1, 1)$, $(8, 2, 1)$, $(7, 3, 1)$, $(7, 2, 2)$, $(6, 3, 2)$, $(5, 3, 3)$. We note that there are rank-3 partitions of 11 into which the partition $(5, 2, 1)$ cannot be packed; clearly, $(4, 4, 3)$ is such.

MODEL 5.9. Suppose we have m batches of inquiries for distinguishing memory. To the sizes of the inquiries in the jth batch correspond the parts of the partitions $(k_1^{(j)}, \ldots, k_{t_j}^{(j)})$, $j = 1, \ldots, m$. Suppose we are also given a state of free computer memory, interpreted as a partition $(n_1, \ldots, n_r) \vdash n$, which, in accordance with some chosen criterion (e.g., the worst, from the point of view of distributing the inquiries over it), characterizes the external fragmentation in the process of functioning of the computer of the automatic control system. Then, in accordance with the definitions in Theorem 2.3 we can prove that the quantity

$$\max_{1\leq j\leq m} n(k_1,\ldots,k_t;n_2,\ldots,n_r) = \max_{1\leq j\leq m}\left(\max_{1\leq i\leq t_j}\left(\sum_{v=1}^{i}k_v^{(j)}+\sum_{l=2}^{r}\min(n_l,k_i^{(j)}-1)\right)\right),$$

where $k_1^{(j)}\geq\cdots\geq k_{t_j}^{(j)}$, is this minimal total size of r fragments of continuous address space of free memory of sizes $p_1,\ldots,p_r$ with $p_i\leq n_i$ $(i=2,\ldots,r)$, respectively. And these sizes p_i $(i=1,\ldots,r)$ guarantee the simultaneous satisfaction of all inquiries of any jth batch $(1\leq j\leq m)$. In this model the algorithm for distributing inquiries can be chosen similarly as in model 5.8.

MODEL 5.10. Suppose the set of all possible states of free computer memory of an automatic control system is interpreted as the set of number partitions $(n_1^{(i)},\ldots,n_{v_i}^{(i)})$, $i=1,\ldots,l$, where any part $n_j^{(i)}$ of the ith partition corresponds to the jth continuous sector of address space of free memory, and v_i is the amount of continuous sectors of address space of free memory. For each ith state of free memory we know the set of batches of inquiries, each of which may require, in the course of functioning of the automatic control system, simultaneous satisfaction of all its inquiries, given the ith state of free computer memory. Suppose that to this set of batches of inquiries for the ith state of free memory there corresponds a set of partitions $(k_1^{(j)},\ldots,k_{t_j}^{(j)})^{(i)}$, where the rank t_j denotes the amount of inquiries in batch j $(i=1,\ldots,l; j=1,\ldots,m_l)$. Then the definitions in Theorem 2.3 imply that, in accordance with model 5.9, the required size of operation memory is the quantity

$$\max_{1\leq j\leq l}\left(\max_{1\leq j\leq m_i} n\left((k_1^{(j)},\ldots,k_{t_j}^{(j)})^{(i)};n_2^{(i)},\ldots,n_{v_i}^{(i)}\right)\right) =$$

$$= \max_{1\leq j\leq l}\left(\max_{1\leq u\leq m_i}\left(\sum_{u=1}^{i}k_u^{(j)}+\sum_{l=2}^{v_i}\min(n_l^{(i)},k_i^{(j)}-1)\right)\right),$$

where $k_1^{(j)}\geq\cdots\geq k_{t_j}^{(j)}$.

We may distinguish two peculiarities of the combinatorial models presented here, which essentially simplify the study of the process of dynamic computer memory allocation:

- the interpretation of the states of free (occupied) memory by number partitions on the one hand makes it possible not to take into account a large part of the states of computer address space, and on the other hand this characterization of memory states includes all necessary parameters to sufficiently well reflect the influence of external fragmentation;

- in the models the time that computer memory is in some or other fragmented state is not taken into account; the characterization using unordered number partitions makes it only possible to assert the fact that memory is in a state admissible for it.

However, the use of these additional features in combinatorial models requires yet the solution of the problem of finding a priori information regarding the functioning of the automatic control system, and in particular regarding the functioning of the operation computer memory. The possibility of and methods for obtaining it have a substantial influence on the efficiency of using combinatorial models in the process of investigation. The matter is that the looked for a priori data is input information for the class of combinatorial models under consideration, so that the completeness and degree of exactness of this data in a finite time span influence the results of the investigation in the large.

The use of combinatorial models for solving the problem of finding an upper bound for the necessary operation computer memory in an automatic control system supposes, first of all, a determination of the main factors under whose influence a state of operation computer memory changes. The reasonings given earlier make it possible to say that the sizes of inquiries on distinguishing operation memory are one such factor. But the sizes of inquiries, in turn, depend on the sizes of the informational objects (programs and array data) forming the software of the automatic control system. Consequently, to obtain a priori information on the functioning of operation computer memory, we must analyze the data concerning the software of the automatic control system.

5. The use of combinatorial models for estimating the necessary size of operation computer memory in an automatic control system

Each function to be executed in the automatic control system (information processing, all possible calculations, accumulation, renewal, and re-allocation of data, etc.) can be realized by various program complexes or software of the automatic control system. Using the complexity specifics of elaboration of software of the automatic control system, and also the aim of creating the system, related to the automation of a process of information processing in some or other concrete domain of knowledge, we can distinguish two main properties of software of automatic control systems:

- modularity of construction of software, making it possible to state in a single manner the requirement on the size of free memory needed for realizing some concrete function (task) of the automatic control system;
- functional closedness of the construction of software of the automatic control system, i.e. the determination of finitely many program modules, and the naming of them, realizing every concrete task, and also of finitely many functions to be executed, which during a long time span of exploiting the system (between moments of updating it) remain unchanged.

The presence of such properties makes it possible to give the following definition of a software structure for an automatic control system.

DEFINITION 5.2. A *software structure of an automatic control system* is a description of a family of functionally closed linear sequences of program and informational modules or informational objects, containing data regarding the maximal sizes of operation computer memory required for executing or loading each module and data regarding the successive execution (use) of these modules for each job of the automatic control system.

At first glance this conception of job execution in automatic control systems may seem erroneous. In fact, many jobs in a system are realized, depending on the input data, by a 'branched' and not linear sequence of program modules. Moreover, each program module can, in the process of its execution, be a source of inquiries on distinguishing memory needed by it to allocate intermediate or output data. However, the appearance of such inquiries in a process of job execution by an automatic control system has no influence on the nature of the sequence of executing the program modules for realizing a concrete function of information processing. Moreover, working out the names of jobs in the automatic control system taking into account input information during a finite time span, leads to an interpretation of them precisely by linear sequences of program and information modules.

The possibility of such a representation of the software structure of an automatic control system is of great importance in the use of the combinatorial approach to studies of the process of functioning of the system. Data regarding the software structure give already at the stage of designing the automatic control system a priori information necessary for applying combinatorial models and for obtaining bounds for the values of the studied parameters of functioning of the system. The degree of exactness of data regarding the software structure is determined by the degree of detail in the design of the automatic control system or by the design stage at which this information is obtained. At different design stages, when the algorithms implementing functions of the automatic control system have not yet been determined in detail, such data may serve as bounds on the necessary size of operation memory for implementing each function of the automatic control system.

It is important to note that the combinatorial models 5.4–5.7 make it possible to obtain a theoretically substantiated upper bound for the size of operation computer memory of the automatic control system when orienting data on the necessary sizes of operation memory appear and in the presence of information regarding the functional meaning of each terminal of the automatic control system. Moreover, the models 5.3–5.7 make it possible to obtain analytic dependence of the bound for the necessary operation computer memory of the automatic control system on such characteristics as parameters of the software structure, the amount and functional meaning of terminals of the systems, etc.

Consider the example of using combinatorial models for finding upper bounds for the necessary size of operation computer memory at various stages of design of the automatic control system, i.e. when only orienting data regarding the necessary sizes of memory for realizing each job from a system $z \in Z$ is available. This data may be given as a list of values of the sizes. Certain jobs of the automatic control system may be given identical orienting sizes of necessary memory in the design, hence in the list there may appear identical entries. This list cannot be regarded as a set anymore, since this would contradict the basic property of elements joined by the notion of a 'set'. A list having identical entries is a multiset (see Chapter 1). However, in our investigations we will more conveniently regard it as a number permutation. On the one hand, this does not contradict with the definition of partition, on the other hand the use of the notion of set of number partitions does not contradict the rule for constructing the set $W(Q)$, expounded in model 5.3.

Suppose the partition $(k_1, \ldots, k_l) \vdash k$ corresponds to initial data regarding memory requirements, while its parts k_i correspond to given orienting memory sizes necessary for realizing the ith job in the automatic control system. From the initial data regarding the functional meaning of each terminal of the automatic control system, used in model 5.3, we construct the set $W(Q) \ni (w, \bar{w})$. We denote by $|w|$

and $|\bar{w}|$ the amount of jobs represented by w and $\bar{w}$, respectively. By replacing each job in w by the necessary size of memory for realizing it, taken from $(k_1, \ldots, k_l) \vdash k$, we obtain a partition $p_r = p(k_{i_1}, \ldots, k_{i_r})$ of rank $r = |w|$, $1 \leq i \leq l$, $1 \leq r \leq f$. Thus, in accordance with the definition of w in model 5.3, p_r characterizes an admissible state of computer memory of the automatic control system. The parts of this partition correspond to the sizes of occupied sectors of memory, and the rank characterizes the maximally possible 'scattering' of free memory in the presence of r continuous occupied sectors of address space in it.

In this way we also put corresponding memory sizes in place of jobs in $\bar{w}$. A a result we obtain a partition $\bar{p}_t = (\bar{k}_{j_1}, \ldots, \bar{k}_{j_t})$ of rank $t = |\bar{w}| = f - r$, $1 \leq j \leq l$, which determines the maximally admissible number of inquiries in a batch and the sizes $\bar{k}_{j_m}$ ($1 \leq m \leq t$) of these inquiries. This also characterizes the possible flow of inquiries on distinguishing memory for the fixed state of computer memory represented by p_r. As a result we obtain a pair of number partitions $(p_r, \bar{p}_t)$, corresponding to the element $(w, \bar{w}) \in W(Q)$, characterizing an admissible situation of computer memory in the functioning of the automatic control system and, moreover, containing all information necessary for estimating, in accordance with models 5.5–5.7, the necessary size of operation computer memory. Putting in this way each pair $(w, \bar{w}) \in W(Q)$ into correspondence with the pair of partitions $(p_r, \bar{p}_t)$, we obtain a set $B(Q)$ of pairs of partitions which are necessary a priori information for giving an upper bound for the necessary size of operation computer memory of the automatic control system under investigation.

We can distinguish two basic, and at the present design stage common, properties of the parameters of the partitions forming pairs $(p_r, \bar{p}_t) \in B(Q)$; to wit:

- the ranks of the partitions in a pair always satisfy the equality $r + t = f$;
- the sum of the numbers from which the partitions $p_r, \bar{p}_t$ are obtained always satisfies the inequality

$$\sum_{j=1}^{r} k_{i_j} + \sum_{m=1}^{t} \bar{k}_{i_m} \leq k'f, \qquad \text{where } k' = \max_{1 \leq i \leq l} k_i,$$

$k_i \in (k_1, \ldots, k_l)$ and $k_{i_j}, \bar{k}_{i_m}$ are the parts of the partitions $p_r, \bar{p}_t$, respectively.

Analytic expressions for obtaining upper bounds of the necessary size of operation computer memory, in dependence of the implemented method of allocating inquiries, can be obtained on the basis of the corresponding models. Suppose that in the automatic control system under investigation we have implemented batch allocation of inquiries on distinguishing memory. For the computer of this automatic control system we can obtain an upper bound for the necessary size of operation memory by using model 5.6; in fact, to each state of operation computer memory corresponds a unique batch of inquiries on distinguishing memory. Using this fact the required quantity V is defined by the expression

$$V = \max_{(p_r, \bar{p}_t) \in B(Q)} \left(\max_{1 \leq m \leq t} \sum_{j=1}^{m} \bar{k}_j + (\bar{k}_m - 1)r + \sum_{i=1}^{r} k_i \right),$$

for $\bar{k}_1 \geq \cdots \geq \bar{k}_t$, where $\bar{k}_j, k_i$ are the parts of the partitions $\bar{p}_t, p_r$, respectively.

For the computer of an automatic control system in which individual allocation of inquiries on distinguishing memory is implemented, we use model 5.7 to estimate the necessary size of operation memory. Taking into account the same peculiarities of the initial data, represented by the set $B(Q)$, the calculation of a bound is a particular case of that of the expression in the model 5.7, and is defined by

$$V' = \max_{(p_r, \bar{p}_t) \in B(Q)} \left(\sum_{j=1}^{t} \bar{k}_j + \left(\max_{1 \le m \le t} \bar{k}_m - 1 \right) r + \sum_{i=1}^{r} k_i \right),$$

where $k_i, \bar{k}_j$ are the parts of the partitions $p_r, \bar{p}_t$, respectively.

Clearly, $V \le V'$, if V and V' are computed for one and the same set $B(Q)$. This relation follows from the rule of constructing the set $W(Q)$, which, in turn, determines the collection of elements $B(Q)$. However, if distinct subsets of $B(Q)$ are taken to compute V and V', then this relation may become different.

It is not difficult to note that the amount of calculations in estimating the size of operation computer memory using the expressions given above is mainly determined by the amount of terminals served by the automatic control system, their functional meaning, and the presence of data regarding software implementing jobs or functions of the automatic control system. The rule for constructing the set $B(Q)$ allows us to determine the amount of calculations necessary to estimate the size of operation computer memory at the present design stage of the automatic control system. The amount of operations to be performed in the given case is defined by the quantity $|W(Q)| = |B(Q)| = 2^f \cdot \prod_{i=1}^{f} |Q_i|$, where the operation is taken to be the calculation of the expression defining V or V' for a single element $(p_r, \bar{p}_t) \in B(Q)$. When data regarding the software structure implementing jobs of the automatic control system emerges, the amount of calculations increases sharply. Let $q = (k_1^{(i)}, \ldots, k_{t_i}^{(i)})$ be a partition, interpreted as the sizes and number of programs involving the necessary array data implementing the ith job in the automatic control system. Using model 5.3, in this case to each job of the automatic control system will correspond not one number (a bound on the necessary size of memory for implementing the job), but t_i numbers (the parts of the partition q). Assume that under execution of a job by the automatic control system the computer memory contains only one sequence of programs implementing this job, i.e. that after execution the programs empty the operation memory occupied. Then, in accordance with the models 5.3, 5.5–5.7, the initial data for the calculation require the execution of $2^f \cdot \prod_{i=1}^{f} R(Q_i)$ operations, where $R(Q_i)$ is the sum of the ranks of the partitions characterizing the software structure implementing the jobs intended for initializing the ith terminal of the automatic control system. Clearly, $\prod_{i=1}^{f} R(Q_i) \gg \prod_{i=1}^{f} |Q_i|$ for $t_i \gg 1$. As already proved above, in studies using combinatorial models, reduction of the enumeration of states of automatic control system elements under investigation is made possible by results of extremal combinatorial problems. In this case such a result reads as follows.

THEOREM 5.2. *Let $p = (k_1, \ldots, k_r) \vdash k$, and let Q be the set of all partitions $q = (k_j)_{j \in B}$, where $B \subset 2^{[r]}$, i.e.*

$$Q = \bigcup_{B \subset 2^{[r]}} (k_j)_{j \in B}.$$

Let $r(p)$ be the rank of the partition p, $(p - q)$ the partition obtained from p by deleting certain parts constituting q, and let $|(p - q)|$ be the sum of the parts of the *partition $(p - q)$. If $k_1 \geq \cdots \geq k_r$, then*

$$\max_{q \in Q}(n(q; r(p - q) + 1) + |p - q|) = k + (k_1 - 1)(r - 1).$$

PROOF. We first prove the upper bound. Substitution of $n(q; r(p - q))$ gives (in accordance with the principle of complete allocation, Theorem 2.2)

$$\max_{q \in Q}(n(q; r(p - q) + 1) + |p - q|) =$$

$$= \max_{\substack{q \in Q \\ q=(q_1,\ldots,q_t) \\ q-1 \geq \cdots \geq q_t}} \left(\max_{1 \leq i \leq t} \left(\sum_{j=1}^{i} q_j + (q_i - 1)(r(p) - r(q)) \right) + \sum_{q_i \in (p-q)} q_i \right).$$

Since $\sum_{q_i \in (p-q)} q_i + \sum_{j=1}^{i} q_j \leq |p|$, $1 \leq i \leq t$, the following inequality holds:

$$\max_{\substack{q \in Q \\ q=(q_1,\ldots,q_t) \\ q-1 \geq \cdots \geq q_t}} \left(\max_{1 \leq i \leq t} \left(\sum_{j=1}^{i} q_j + (q_i - 1)(r(p) - r(q)) \right) + \sum_{q_i \in (p-q)} q_i \right) \leq$$

$$\leq |p| + \max_{\substack{q \in Q \\ q=(q_1,\ldots,q_t) \\ q-1 \geq \cdots \geq q_t}} \left(\max_{1 \leq i \leq t}(q_i - 1)(r(p) - r(q)) \right).$$

But

$$\max_{\substack{q \in Q \\ q=(q_1,\ldots,q_t) \\ q-1 \geq \cdots \geq q_t}} \left(\max_{1 \leq i \leq t}(q_i - 1)(r(p) - r(q)) \right) \leq (q_1 - 1)(r - 1),$$

so that

$$|p| + \max_{\substack{q \in Q \\ q=(q_1,\ldots,q_t) \\ q-1 \geq \cdots \geq q_t}} \left(\max_{1 \leq i \leq t}(q_i - 1)(r(p) - r(q)) \right) \leq$$

$$\leq |p| + (q_1 - 1)(r - 1) \leq k + (k_1 - 1)(r - 1).$$

It remains to prove that the bound is always realized. In fact, for $q = (k_1)$ we have

$$\max_{q \in Q}(n(q; r(p - q) + 1) + |p - q|) \geq n((k_1); r(p)) + |p - (k_1)| =$$

$$= k_1 + (k_1 - 1)(r - 1) + k - k_1 = k + (k_1 - 1)(r - 1).$$

This proves the theorem. □

This extremal result substantially reduces the investigated amount of initial data when computing upper bounds for the necessary size of operation computer memory. In the general case, for each state of the automatic control system, characterized in model 5.3 by simultaneous satisfaction of jobs of a complete family, one assumes the construction of a set of pairs $(q_i, \bar{q}_i)$, $q_i \in Q(q)$, $\bar{q}_i \in Q(q)$, which characterize the states of the system existing before its transition to the state $q \in Q$. However, in transition to actual requirements on memory, characterized by pairs of partitions $(p_r, \bar{p}_t) \subset B(Q)$, the set $B(Q)$ can be made substantially smaller. Let $B(q) \subset B(Q)$ be the set of pairs of partitions $(p_r, \bar{p}_t)$ corresponding to $(q_i, \bar{q}_i)$. In turn, the latter are given by a bijective correspondence $\phi: Q(q) \rightarrow Q(q)$, where $\phi(q_i) = \bar{q}_i = q \setminus q_i$, $q \in Q$. Clearly, the parameters of every pair $(p_r, \bar{p}_t) \subset B(q)$ satisfy the estimates

$$r + t = f; \qquad \sum_{j=1}^{r} k_{i_j} + \sum_{m=1}^{t} \bar{k}_{i_m} = \sum_{i=1}^{f} k_i, \qquad 1 \le i \le f,$$

where k_{i_j} are the parts of the partition p_r, $1 \le j \le r$, $\bar{k}_{i_m}$ are the parts of the partition $\bar{p}_t$, $1 \le m \le t$ and k_i are the parts of the rank-f partition which is the list of required operation memory for implementing the jobs in $q \in Q$. Then it is easy to prove that, in accordance with the results of Theorem 5.2:

$$\max_{(p_r, \bar{p}_t) \subset B(q)} \left(\max_{1 \le m \le t} \sum_{v=1}^{m} \bar{k}_{i_v} + (\bar{k}_{i_m} - 1) + \sum_{j=1}^{r} k_{i_j} \right) = \sum_{i=1}^{f} k_i + \left(\max_{i \le i \le f} k_i - 1 \right) r.$$

Thus, in the model 5.3, for estimating the size of operation computer memory necessary for simultaneously satisfying the jobs in $q \in Q$, for each $q \in Q$, we need not study all 2^f states preceding the transition of the system to state q. It suffices to consider one such state, characterized by a pair of partitions (p_r, p_t') such that

$$p_t' = (k_{i_1}), \qquad p_r = (k_{i_2}, \ldots, k_{i_r}),$$

$k_{i_1} = \max_{1 \le i \le f} k_i$, $r = f - 1$. Hence, to compute an upper bound for the necessary size of operation computer memory, as initial data we may take a set of partitions $P(Q)$ each element $p(q) = (k_1, \ldots, k_f)$ of which is the list of memory sizes required for implementing the jobs from the complete family $q \in Q$ corresponding to it. Using Theorem 5.2, to determine V' it suffices to compute

$$V(Q) = \max_{\substack{p(q) \in P(Q) \\ q \in Q}} \left(\sum_{i=1}^{f} k_i + \left(\max_{1 \le i \le f} k_i - 1 \right) r \right).$$

The quantity $V(Q)$ can be used as upper bound for the necessary size of operation computer memory in an automatic control system under both batch and individual methods of satisfying inquiries on memory. It is easy to prove that here the required amount of calculations, as in the previous cases, is determined by the set of initial data. In fact,

$$|P(Q)| = \prod_{i=1}^{f} |Q_i| \ll 2^f \cdot \prod_{i=1}^{f} |Q_i|.$$

We note that the parts of each partition $p(q)$ can be chosen, in accordance with the rule given in model 5.3, from one and the same list of values $(k_1, \ldots, k_l)$, where k_i corresponds to the necessary size of continuous address space of operation memory which is needed for implementing the ith job of the system and l is the number of possible data that can be implemented by the automatic control system. Using the fact that in finding $V(Q)$ the quantity $r = \text{const}$, enumeration of the elements of $P(Q)$ in the determination of $V(Q)$ is completely excluded by the following lemma.

LEMMA 5.1. Let $\lambda(k_{i_1}, \ldots, k_{i_r}) = \sum_{j=1}^{r} k_{i_j} + \left(\max_{1 \le i \le r} k_{i_j} - 1 \right)$, $(k_{i_1}, \ldots, k_{i_r}) \in R$, the set of all r-pairings of elements of the set $N = \{k_{i_1}, \ldots, k_{i_r}\}$, $k_i > 0$, $k_1 \ge$

$\cdots \geq k_l$, $l > r$. Then

$$\max_{(k_{i_1}, \ldots, k_{i_r}) \in R} \lambda(k_{i_1}, \ldots, k_{i_r}) = \sum_{i=1}^{r} k_i + (k_1 - 1)r.$$

PROOF. Using the rule for constructing R, we determine the quantity

$$\max_{(k_{i_1}, \ldots, k_{i_r}) \in R} \left(\max_{1 \leq i \leq r} k_{i_j} \right) \leq \max_{1 \leq i \leq l} k_i = k_1, \qquad k_i \in \mathbf{N}.$$

But $\max_{(k_{i_1}, \ldots, k_{i_r}) \in R} \sum_{j=1}^{r} k_{i_j} \leq \sum_{i=1}^{r} k_i$, i.e. the maximum is determined by the sum of maximal elements of N, and in fact by the elements $(k_1, \ldots, k_r)$. By definition, the elements of R are all possible r-pairings of elements of N, i.e. $(k_1, \ldots, k_r) \in R$, as required. □

The result of Lemma 5.1 makes it possible to assume that the quantity $V(Q)$ is determined by a partition $p'(q) \in P(Q)$ with maximal parts in comparison to other elements of $P(Q)$. We will call $p'(q)$ the *extremal size of the software structure* for the concrete distribution of functions over terminals of the automatic control system. An extremal section of a software structure has important value in studying and optimizing parameters characterizing the process of computer memory allocation. To find $p'(q)$ without enumeration of the elements of $P(Q)$ makes it possible to give a simple construction rule for $P(Q)$, which uses the ordered list of a priori data regarding the sizes of inquiries on memory in the automatic control system. To sum up th e results obtained, we determine the order of actions in estimating from above the necessary size of operation computer memory in an automatic control system.

6. The order of calculating a bound for the necessary size of operation computer memory in an automatic control system

First it is necessary to analyze the existing data regarding the software structure implementing jobs in the automatic control system, and to tie them to data regarding the functional meaning of terminals of the automatic control system. To this end we represent data regarding the necessary sizes of memory needed for implementing jobs in the automatic control system by a list of values $N = (k_1, \ldots, k_l)$. Depending on the stage of design under consideration, the elements k_i ($1 \leq i \leq l$) of this list may be data regarding oriented memory sizes for implementing jobs, or data regarding the sizes of all programs and their arrays which make up the software of the automatic control system. Let N be the list of values of memory sizes needed for implementing the programs making up the software of the automatic control system. We number the elements of this list from 1 to l, and order them such that $k_1 \geq \cdots \geq k_l$.

To tie the initial data regarding the software structure to the functional meaning of terminals of the system, we must construct the set Q_i ($1 \leq i \leq f$) whose elements are jobs intended to be initialized by the ith terminal. From N and Q_i we then obtain, using the map $\phi \colon N \to Q_i$ such that

$$\phi^{-1}(z) = \{k_{i_j} : z = \phi(k_{i_j}), \, k_{i_j} \in N, \, z \in Q\} \qquad (1 \leq i \leq f),$$

a list $Q_i(N)$ of memory sizes required for implementing the jobs making up the set Q_i. The elements of $Q_i(N)$, as well as those of N, are ordered by decreasing value.

Using the lists $Q_i(N)$ $(1 \leq i \leq f)$ we determine the extremal section of the software structure of the automatic control system, $p'(q)$, using the following rule:

$$p'(q) = (k_{i_j} : k_{i_j} \in Q_m(N), \, j = \min(j \in L_m)), \qquad 1 \leq m \leq f,$$

where f is the amount of terminals to be served by the automatic control system, and L_m is the set of indices of the elements in $Q_m(N)$, $L_m \subset \{1, \ldots, l\}$.

The calculation of an upper bound for the necessary size of operation computer memory in the automatic control system is, in accordance with Lemma 5.1, conducted on the elements of $p'(q)$, in correspondence with the following expression:

$$V(Q) = \sum_{i=1}^{f} k_{i_j} + (k_{i_1} - 1)f,$$

where $k_{i_j} \in p'(q)$, $1 \leq i \leq l$.

Thus, the successive solution of easy extremal combinatorial problems concerning packability in the set of number partitions, interpreted as the possible states of the system under investigation, completely excludes the enumeration of all its elements in finding the value of a functional defined on the parameters of the elements of this set.

The rule for finding an extremal section of the software structure of the automatic control system has polynomial computational complexity. This guarantees a relatively fast solution of this problem on a computer, for software structures of practically any complexity. The selection of a sector of operation memory in correspondence with $V(Q)$ is guaranteed by the use of the simplest method for controlling memory allocation, the software for whose realization are given by that unique first-fit memory allocation algorithm. The quantity $V(Q)$ is an upper bound for the necessary size of computer memory, therefore, of course, it exceeds the real requirements on memory in implementing some software structure and collection of programming tools of the automatic control system. We can determine the efficiency of applying some extremal result regarding the packability of number partitions to compute an upper bound of the necessary size of operation computer memory. To this end, next to the comparison of the extremal results obtained with each other, it is useful to have some 'absolute' value of the parameters under investigation. The method for comparing various extremal results regarding packability of number partitions which is most simple to state is that of comparison by the results of solving the packability problem for a pair of partitions definitely known to be packable. The simplest formulation of the problem in this case is to verify 'self-packability'

$$(k_1, \ldots, k_t) \subseteq (k_1, \ldots, k_t).$$

This formulation shows, for each extremal result giving a bound for the necessary size of free memory, how much memory will guarantee the simultaneous allocation of inquiries of sizes $(k_1, \ldots, k_t)$ in fragmented address space of free memory with sizes $(k_1, \ldots, k_t)$ of free sectors. Using Dirichlet's principle (within the model 5.7) for determining the necessary size of operation computer memory, in accordance with the approach chosen we obtain the bound $tk - t + 1$.

In the models 5.4–5.7 the method of complete allocation leads to the bound

$$n(k_1, \ldots, k_t; t) = \max_{1 \le i \le t} \left(\sum_{j=1}^{i} k_j + (k_i - 1)(t - 1) \right)$$

for the necessary size of operation computer memory. By the upper bound in (2.27) the least possible bound for the necessary size of operation computer memory is $k \cdot 1.5819$, since

$$m(k, t, t) < \frac{kt}{t^t - (t-1)^t} \le \frac{k}{1 - e^{-1}} = k \cdot 1.5819.$$

This result plainly shows to what amount the principle of complete allocation is more efficient than Dirichlet's principle.

The use of Theorem 2.3 to determine the necessary size of operation computer memory gives the following bounds.

PROPOSITION 5.3. *Let*

$$M(k, t) = \max_{\substack{(k_1, \ldots, k_t) \vdash k \\ k_1 \ge \cdots \ge k_t}} n(k_1, \ldots, k_t; k_2, \ldots, k_t).$$

Then

$$k + (t - 1) \left(\left[\tfrac{k}{t} \right] - 1 \right) \le M(k, t) \le 2k - \left[\tfrac{k}{t} \right] - t + 1.$$

PROOF. The following equation holds:

$$n(k_1, \ldots, k_t; k_2, \ldots, k_t) = k + \max_{2 \le i \le t} (k_i - 1)(i - 1).$$

In fact,

$$n(k_1, \ldots, k_t; k_2, \ldots, k_t) = \max_{1 \le i \le t} \left(\sum_{j=1}^{i} k_j + \sum_{l=2}^{t} \min(k_l, k_i - 1) \right) =$$

$$= \max_{1 \le i \le t} \left(\sum_{j=1}^{i} k_j + \sum_{j=i+1}^{t} k_l + (i - 1)(k_i - 1) \right) = k + \max_{2 \le i \le t} (k_i - 1)(i - 1).$$

The lower bound is realized by the most uniform partition: $(]k/t[, \ldots,]k/t[) \vdash k$, hence the inequality $k_i \ge [k/t]$ holds for the maximizing partition $(k_1, \ldots, k_t) \vdash k$ and its maximizing index i.

We now prove the upper bound by the method of contradiction. Assume that $(i - 1)k_i - i + 1 > k - [k/t] - t + 1$. Then

$$(i - 1)k_i - i + 1 > k - \left[\tfrac{k}{t} \right] - t + 1 = k_1 + \cdots + k_t - \left[\tfrac{k}{t} \right] - t + 1 \ge$$

$$\ge (i - 1)k_i + k_i + k_{i+1} + \cdots + k_t - \left[\tfrac{k}{t} \right] - t + 1 \ge$$

$$\ge (i - 1)k_i + k_i + t - i - \left[\tfrac{k}{t} \right] - t + 1 = (i - 1)k_i + k_i - \left[\tfrac{k}{t} \right] + 1,$$

or $k_i < [k/t]$, contradicting the previous remark. $\square$

The given system of bounds, obtained by using results of solving extremal combinatorial problems, plainly demonstrate the fact that with increasing a priori information regarding the functioning of the elements under investigation the results become substantially more refined, i.e. the efficiency of applying combinatorial models in these investigations increases. To stress this it suffices to compare the bounds given above.

It is easy to prove that if the size of operation computer memory is chosen in agreement with models 5.4–5.6, then in the process of functioning of the automatic control system the efficiency of using memory does not exceed the value

$$\frac{\sum_{j=1}^{f} k_{i_j}}{\sum_{j=1}^{f} k_{i_j} + (k_{i_1} - 1)f} \cdot 100\%.$$

Consequently, for a software structure of the automatic control system in which the elements of an extremal section guarantee that a condition

$$\sum_{j=1}^{f} k_{i_j} > (k_{i_1} - 1)f$$

holds, the efficiency of using operation memory will exceed 50%. It is clear from the last inequality that increase of the actual required size of memory, defined by the very quantity $(k_{i-1} - 1)f$, uses (in accordance with the definitions in the models 5.5–5.7) maximal influence of external fragmentation on the process of computer memory allocation. The appearance of external fragmentation is given here under the assumption that having f occupied sectors in memory leads to the formation of an $(f + 1)$st free sector, while the sizes of these free sectors are one less than the maximal possible size of an inquiry on memory which may arise during the functioning of the system. This assumption on the influence of external fragmentation of memory reflects to a sufficient extent the actual process of functioning of a computer's memory. The '50-percent rule' proved in [41] testifies of this. However, this rule characterizes a steady-state regime of functioning of memory, or, according to the definitions in [41], an equilibrium state, in which there are on the average n occupied sectors in the system. Moreover, the 50-percent rule uses the notion of probability, which makes it impossible to use it as a characteristic of the phenomenon of fragmentation in calculating the size of operation computer memory which would guarantee satisfaction of every inquiry on memory without using the means of reorganization and re-allocation.

The results given show that the required size of computer memory of an automatic control system, necessary for implementing the investigated software structure, is mainly determined by the elements of an extremal sector of the structure. Using the random nature of the process of functioning of memory, we suppose that precisely these elements determine the maximal influence of external fragmentation. Consequently, by investigating the whole set of states of address memory space, being determined by the various situations of simultaneous execution of elements of an extremal sector of the software structure, we can substantially refine the bound $V(Q)$. These investigations can be conducted by using both the method of imitation modeling, as well as the method of extremal combinatorial analysis. The most important is that these investigations need not be conducted on the whole software structure of the automatic control system, but only on the elements of an extremal sector of it.

The quantity $V(Q)$ and the extremal sector of the software structure are of great importance for work scheduling during the design and extension of functions of the automatic control system. In fact, for the scheduling of the computational process here considered the quantity $V(Q)$ does not depend on the amount of programs by which some job of the automatic control system is executed. Therefore the choice of operation memory size in accordance with $V(Q)$ makes it possible to unrestrictedly augment functions of the automatic control system without increasing the size of operation computer memory. This guarantees that the following single condition is fulfilled when extending functions of the automatic control system:

> the software realizing additionally introduced functions need not bring changes in the extremal sector of the software structure of the automatic control system.

An extremal sector of the software structure can be found already at an early stage in the design of the automatic control system, using the mathematical apparatus expounded here. Consequently, data regarding the extremal sector can be noted in documentation on the design and implementation of software. Taking notice of these data as restrictions on the admissible sizes of program and informational modules allows one to fully exclude, in the creation of the system, the solution of problems related with selecting the size of operation computer memory of the automatic control system.

Despite the fact that the quantities V, V', and $V(Q)$ are upper bounds for the necessary size of computer memory (which are even not applicable in realizing certain automatic control systems), they express an analytical dependence of such characteristics of automatic control systems as peculiarities of the software structure, the amount of terminals to be served by the automatic control system, the amount of functions to be implemented by it, and the functional meaning of the terminals. Consequently, the models 5.3–5.7 and the solutions of the extremal combinatorial problems obtained may be regarded as a mathematical apparatus for studying and optimizing the process of control of computer memory, the software structure, and the distribution of functions between terminals of the system.

Moreover, the considered models of restrictions on controlling the computational process, prescribing the finishing of a program by emptying the memory occupied by it, in no way restrict the scope of the investigations conducted using this apparatus. When mapping the set of initial data (model 5.7) onto actual data concerning the software structure using the formal apparatus proposed, one may use a number of additional models, making it possible to study the working of programs with array data, the presence of re-entering and nonre-entering programs in the software structure of automatic control systems, and the implementation of other methods of controlling the course of the computational process. Consequently we have the possibility to study, by these models, all above given peculiarities of functioning of automatic control systems.

For example, by correcting the software structure with regard to the sizes of programs and arrays, and also by shifting certain programs into the stock of residing programs, one obtains the possibility of optimizing the structure and of controlling the computational process with regard to resource demands on operation computer memory; one may study the influence of re-entering of a program, for a given allocation of functions between the terminals of the automatic control system, on the necessary size of operation memory, etc. It is here important to note that the proposed approach to investigations allows us to solve two classes of combinatorial problems: problems concerning the determination of the presence of an admissible solution (on the packability of partitions), and problems of constructing theoretically substantiated algorithms for realizing this solution in polynomial time (an algorithm for packing partitions).

In conclusion we note that the expense of processor time on the organization of control of computer memory allocation in an automatic control system with highly dynamical incoming inquiries during functioning, constitutes more than 1/3rd of the overall time for solving functional problems. The choice of the size of operation memory using the proposed approach makes it possible to implement the easiest method for controlling allocation in the system. This reduces the inproductive expense of processor time on the organization of control of allocation of operation or additional computer memory of the automatic control system. The software realizing this organization of control of operation memory is almost exclusively the program of the **first-fit** algorithm. By excluding from the functions of computer software of the automatic control system the tools for controlling the operation memory, and by implementing the **first-fit** algorithm, one can substantially increase the productivity of a system of this class.

7. Deciphering of passwords

In computers and computer networks one uses passwords, i.e. words consisting of the letters, ciphers and signs available on a keyboard. This is done with the purpose of restricting user access to certain programming means. For example, before entering the American Mathematical Society database one must choose the passwords **e-math**.

Assume that we do not know a password necessary for us to enter a program or database. How to guess this unknown password?

One of the methods is given in problem 3.27. However, it turns out that the weighing problem can also be presented differently. Each sign in a password must be taken from a certain set of values (letters, ciphers and signs; e.g. 40 different ones), so we may suppose that each sign is taken from the 40 distinct numbers $1, 2, \ldots, 40$. In fact, letters, ciphers and signs are coded in the computer (in the binary system) just like this. So, the question of determining a password is equivalent to determining a system of independent loads for parallel weighing with as many weights as the word is long. Therefore the most economic system of weights (i.e. numbers) for deciphering a password can be determined from the results of Chapter 2 concerning single-pan weighing on parallel weights. That is, the inequality

$$\sum_{i=1}^{t} k_i \geq n(k_1, \ldots, k_t; r)$$

implies that the fastest sequence for deciphering two-letter passwords is determined by the system

$$k_i \leq 1 + \left[\frac{1}{2} \sum_{j=1}^{i-1} k_j \right], \qquad i = 1, 2, \ldots,$$

and has the form $1, 1, 2, 3, 4, 6, 9, 14, 21, 31, \ldots$.

Bibliography

[1] O.I. Aven, Ya.A. Kogan, *Control of the computational process in a computer*, Energiya, Moscow, 1978. (In Russian.)

[2] O.I. Aven, N.N. Gurin, Ya.A. Kogan, *Quality bounds and optimisation of computational systems*, Nauka, Moscow, 1982. (In Russian.)

[3] S.S. Agayan, A.G. Sarukhanyan, Recurrence formulas for constructing matrices of Williamson type, *Mat. Zametki* **30: 4** (1981), pp. 603–617. (In Russian.)

[4] M. Aigner, *Combinatorial theory*, Springer, 1979.

[5] V.I. Baranov, A combinatorial model for the phenomenon of memory fragmentation, *Programmirovanie* **3** (1978), pp. 46–54. (In Russian.)

[6] V.I. Baranov, An extremal problem on number partitions, *mat. Zametki* **29: 2** (1981), pp. 303–307. (In Russian.)

[7] V.I. Baranov, Application of methods of combinatorial analysis in the design of algorithms for controlling computer memory allocation, *Programmirovanie* **4** (1985), pp. 33–38. (In Russian.)

[8] V.I. Baranov, Application of methods of combinatorial analysis for calculating computer memory size, *Voprosy Kibernet. (Exploitation and use of supercomputers)* (1986), pp. 191–215. (In Russian.)

[9] V.I. Baranov, Conditions for inclusion of partitions in dependence on the number of terms, *Material Vsesoyuzn. Seminara po Diskretn. Mat. i ee Prilozhen.* MGU, Moscow, (1986), pp. 62–65. (In Russian.)

[10] V.I. Baranov, Combinatorial models for choosing computer memory size, *Programmirovanie* **2** (1987), pp. 91–102. (In Russian.)

[11] V.I. Baranov, Application of combinatorial models to determine requirements on operating memeory size, *Programmirovanie* **6** (1987), pp. 69–80. (In Russian.)

[12] C.G. Bachet, *Games and problems, based on mathematics*, Sint Petersburg, 1877. (Translated from the French.)

[13] G. Birkhoff, *Lattice theory*, Amer. Math. Soc., 1967.

[14] M. Blekman, *Design of real-time systems*, Mir, Moscow, 1977. (In Russian.)

[15] N.Ya. Vilenkin, *Combinatorics*, Nauka, Moscow, 1969. (In Russian.)

[16] I.M. Vinogradov, *Fundamentals of number theory*, Nauka, Moscow, 1965. (In Russian.)

[17] G.P. Gavrilov, A.A. Sapozhenko, *Collection of problems on discrete mathematics*, Nauka, Moscow, 1977. (In Russian.)

[18] V.F. Garti, *Best systems of weighted weights*, Sint Petersburg, 1910. (In Russian.)

[19] V.O. Groppen, *Models and algorithms of combinatorial programming*, Rostov Univ., Rostov n/D, 1983. (In Russian.)

[20] R. Graham, *Initials of Ramsey theory*, Mir, Moscow, 1984. (In Russian.)

[21] M. Geri, D. Johnson, *Computing machines and hard problesm*, Mir, Moscow, 1982. (In Russian.)

[22] E.S. Davydov, *Smallest groups of numbers for forming the natural series*, Sint Petersburg, 1903. (In Russian.)

[23] J. Donovan, *System programming*, McGraw-Hill, 1972.

[24] V.A. Evstegneev, L.S. Mel'nikov, *Problems and exercises on graph theory and combinatorics*, NGU, Novosibirsk, 1981. (In Russian.)

[25] A.L. Ershov, Reduction of the problem of memory allocation in writing programs to the problem of graph vertex coloring, *Dokl. AN SSSR* **142: 4** (1962), pp. 785–787. (In Russian.)

[26] K. Ziegler, *Methods of designing programming systems*, ?.

[27] Selected excerpts of the mathematical work of Leibniz, *UMN* **3: 1(23)** (1948), pp. 165–204. (In Russian.)

[28] P.J. Cameron, J.H. van Lint, *Graph theory, coding theory and block designs*, Cambridge Univ. Press, 1975.

[29] V.M. Karas', Stability of an optimal segmentation of programs, *Programmirovanie* **5** (1987), pp. 75–84. (In Russian.)

[30] D. Katona, Inequalities for the distribution of the length of a sum of random vectors, *Teor. Veroyatn. i ee Primen.* **22: 3** (1977), pp. 466–481. (In Russian.)

[31] D. Katona, A. Kostochka, B. Stechkin, On locally Hamiltonian graphs, *Preprint MIAN VNR (Budapest)* (1982), (In Russian.)

[32] D. Katona, A.V. Kostochka, Ya. Pakh, B.S. Stechkin, Locally Hamiltonian graphs, *Mat. Zametki* **45: 1** (1989), pp. 36–42. (In Russian.)

[33] D. Katona, A.F. Sidorenko, B. Stechkin, On inequalities which hold in the class of all distributions, *Perbyi Vsemirnyi Kongress Obshch. Mat. Stat. i Teor. Veroyatn. im. Bernoulli: Tezis., Nauka, Moscow* **2** (1986), 500. (In Russian.)

[34] D. Katona, B. Stechkin, Combinatorial numbers, geometrical constants, and probabilistic inqualities, *Dokl. AN SSSR* **251: 6** (1980), pp. 1293–1296. (In Russian.)

[35] B.S. Kashin, S.V. Konyagin, On systems of vectors in Hilbert space, *Trudy MIAN* **45: 1** (1989), pp. 36–42. (In Russian.)

[36] A.Ya. Kiruta, A.M. Rybinov, E.B. Yanovskaya, *Optimal choice of allocation in complicated socail-economic problems*, Nauka, Leningrad, 1980. (In Russian.)

[37] A.N. Kolmogorov, S.V. Fomin, *Introduction to the theory of functions and functional analysis*, Nauka, Moscow, 1982. (In Russian.)

[38] K.A. Rybnikov (ed.), *Combinatorial analysis*, Naua, Moscow, 1982. (In Russian.)

[39] G.N. Kopylov, On maximal paths and cycles in graphs, *Dokl. AN SSSR* **234: 1** (1977), pp. 19–21. (In Russian.)

[40] N.A. Krinitskii, G.A. Mironov, *Automated information systems*, Nauka, Moscow, 1982. (In Russian.)

[41] D. Knut, *The art of computer programming*, **1. Fundamental algorithms** Addison-Wesley, 1968.

[42] A. Kaufmann, *Introduction à la combinatorique*, Dunod, 1968.

[43] P. Lancaster, *Theory of matrices*, Acad. Press, 1969.

[44] V.I. Levinshtein, On bounds for packings in n-space, *Dokl. AN SSSR* **245: 6** (1979), pp. 1299–1303. (In Russian.)

[45] V.K. Leont'ev, Discrete extremal problems, *INiT* **16** (1979), pp. 39–101. (In Russian.)

[46] V. Lipskii, *Combinatorics for programmers*, ?.

[47] M.V. Lomonosov, Reasonings on rigidity and thiness of bodies, *Poln. Sobr. Coch. AN SSSR* **2** (1952), pp. 377–410. (In Russian.)

[48] F.G. Enslou (ed.), *Multiprocessor systems and control of computations*, ?.

[49] Yu.V. Matiyasevich, Diophantine sets, *UMN* **27: 5** (1972), pp. 185–222. (In Russian.)

[50] C.H. Papadimitriou, K. Steiglitz, *Combinatorial optimization. Algorithms and complexity*, Prentice-Hall, 1982.

[51] *Enumerative problems of combinatorial analysis*, ?.

[52] M.L. Platonov, *Combinatorial numbers of classes of maps and their application*, Nauka, Moscow, 1979. (In Russian.)

[53] H.J. Ryser, *Combinatorial mathematics*, Wiley, 1963.

[54] E. Reingold, J. Nievergelt, N. Deo, *Combinatorial algorithms. Theory and practice*, Prentice-Hall, 1977.

[55] J. Riordan, *An introduction to combinatorial analysis*, Wiley, 1958.

[56] J. Riordan, *Combinatorial identities*, Wiley, 1968.

[57] K.A. Rybnikov, *Introduction to combinatorial analysis*, MGU, Moscow, 1985. (In Russian.)

[58] T. Saaty, *Optimization in integers and related extremal problems*, McGraw-Hill, 1970.

[59] S.G. Sal'nikov, Locally Ramsey properties of graphs, *Mat. Zametki* **43: 1** 1988. pp. 133–142. (In Russian.)

[60] V.N. Sachkov, *Introduction to the combinatorial methods of discrete mathematics*, Nauka, Moscow, 1982. (In Russian.)

[61] V.N. Sachkov, Classical combinatorial problems, *Mat. Encycl.* **2** (1979), (In Russian.)

[62] V.N. Sachkov, *Combinatorial methods of discrete mathematics*, Nauka, Moscow, 1977. (In Russian.)

[63] V.N. Sachkov, Combinatorial analysis, *Mat. Encycl.* **2** (1979), (In Russian.)

[64] A.F. Sidorenko, Classes of hypergraphs and probabilistic inequalities, *Dokl. AN SSSR* **254: 3** (1980), pp. 540–543. (In Russian.)

[65] A.F. Sidorenko, On the locally Turan property for hypergraphs, *Komb. Anal.* **7** (1986), pp. 146–154. (In Russian.)

[66] A.F. Sidorenko, On the maximal number of edges in a homogeneous hypergraph, not containing forbidden subgraphs, *Mat. Zametki* **41: 3** (1987), pp.

433–455. (In Russian.)

[67] A.F. Sidorenko, On exact values of the Turan number, *Mat. Zametki* **42: 5** (1987), pp. 751–760. (In Russian.)

[68] A.F. Sidorenko, Extremal bounds of probability measures and their combinatorial nature, *Izv. AN SSSR Ser. Mat.* **46: 3** (1982), pp. 535–568. (In Russian.)

[69] A.F. Sidorenko, B.S. Stechkin, On the computation and application of extremal geometrical constants, *Perb. Konferents. po Kombinat. Geometrii i ee Primenen.: Tezis., Batum Univ., Batum* (1985), pp. 59–62. (In Russian.)

[70] A.F. Sidorenko, B.S. Stechkin, On a new class of probability inequalities, *Trudy Mezhdunarodn. Vil'nyussk. Knoferents. po Teorii Veroyatnost. i Mat. Statist.: Tezis. Dokladov, Vil'nyus* **2** (1981), pp. 149–150. (In Russian.)

[71] A.F. Sidorenko, B.S. Stechkin, On a class of extremal geometrical constants and their application, *Mat. Zametki* **45: 3** (1988), pp. ?. (In Russian.)

[72] A.F. Sidorenko, B.S. Stechkin, Extremal geometrical constants, *Mat. Zametki* **29: 5** (1981), pp. 691–709. (In Russian.)

[73] B.S. Stechkin, Asymptotics for local properties of graphs, *Dokl. AN SSSR* **275: 6** (1984), pp. 1320–1323. (In Russian.)

[74] B.S. Stechkin, Binary functions on ordered sets (inversion theorems), *Trudy MIAN* **143** (1977), pp. 178–187. (In Russian.)

[75] B.S. Stechkin, Inclusion of partitions, *Preprint MIAN VNR, Budapest* (1983), (In Russian.)

[76] B.S. Stechkin, locally bipartitie graphs, *Mat. Zametki* **44: 2** (1988), pp. 216–224. (In Russian.)

[77] B.S. Stechkin, Compositions and their use in combinatorial schemes (on acombinatorial formalisation), *Kombinatorn. i Asimptotichesk. Anal., Krasnoyarsk. Univ.* **2** (1977), pp. 44–54. (In Russian.)

[78] B.S. Stechkin, Yamamoto's inequality and compositions, *Mat. Zametki* **19: 1** (1976), pp. 155–160. (In Russian.)

[79] B.S. Stechkin, Some unsolved combinatorial problems, *Zbornik radova Mat. Inst. Belgrad Nov. Ser.* **2 (10)** (1977), pp. 129–137. (In Russian.)

[80] B.S. Stechkin, On the Bachet—Mendeleev problem, *Kvant* **8** (1988), (In Russian.)

[81] B.S. Stechkin, On monotone subsequences in permutations of n natural numbers, *Mat. Zametki* **13: 4** (1973), pp. 511–514. (In Russian.)

[82] B.S. Stechkin, On the fundamnetals of real Möbius theory, *preprint Inst. Fiziki L.V. Kurenskii SO AN SSSR Krasnoyarsk* (1979), (In Russian.)

[83] B.S. Stechkin, Generalised valencies, *Mat. Zametki* **17: 3** (1975), pp. 432–442. (In Russian.)

[84] B.S. Stechkin, The principle of complete arrangement, in: [20], pp.87–96 (Russian translation).

[85] B.S. Stechkin, Inclusion theorems for the Möbius function, *Dokl. AN SSSR* **260: 1** (1981), pp. 40–44. (In Russian.)

[86] B.S. Stechkin, Extremal properties of number partitions, *Dokl. AN SSR* **264: 4** (1982), pp. 833–836. (In Russian.)

[87] B.S. Stechkin, Extremal properties of partitions, in [100], pp. 249–253 (Russian translation).

[88] B.S. Stechkin, P. Frankl, Locally Turan property for k-graphs, *Mat. Zametki* **29: 1** (1981), pp. 83–94. (In Russian.)

[89] V.E. Tarakanov, *Combinatorial problems and $(0,1)$-matrices*, Nauka, Moscow, 1985. (In Russian.)

[90] R. Tirof, *Information processing in control*, Mir, Moscow, 1976. (In Russian.)

[91] R.J. Wilson, *Introduction to graph theory*, Oliver & Boyd, 1972.

[92] J. Fox, *Software and its elaboration*, ?.

[93] G. Hardy, *Divergent series*, Clarendon Press, 1949.

[94] M. Hall, *Combinatorial theory*, Blaisdell, 1967.

[95] D. Tsikriizis, F. Bernstein, *Operating systems*, Mir, Moscow, 1977. (In Russian.)

[96] R. Shennon, *Imitation modelling - art and science*, Mir, Moscow, 1978. (In Russian.)

[97] A.N. Shiryaev, *Probability*, Nauka, Moscow, 1980. (In Russian.)

[98] L. Bic, A.C. Shaw, *The logical design of operating systems*, Prentice-Hall, 1988.

[99] L. Euler, *Introductio in analysis infinitorum*, 1746.

[100] G. Andrews, *Theory of partitions*, Addison-Wesley, 1976.

[101] P. Erdös, J. Spencer, *Probabilistic methods in combinatorics*, Acad. Press, 1974.

[102] I.M. Yaglom, *The problem of the thirteen spheres*, Vishcha Shkola, Kiev, 1975. (In Russian.)

[103] A. Aiello, E. Burattini, A. Massarotti, F. Ventriglla, A poseteriori evaluation of bin packing approximation algorithms, *Discrete Appl. Math.* **2** (1980), pp. 159–161.

[104] Zs. Baranyai, On factorization of the complete uniform hypergraph, *Infinite and Finite Sets* **1** (1975), North-Holland, pp. 91–108.

[105] E.A. Bender, J.R. Goldman, On the applications of Möbius inversion in combinatorial analysis, *Amer. Math. Monthly* **82: 8** (1975), pp. 789–802.

[106] C. Benson, Minimal regular graphs of girth eight and twelve, *Canad. J. Math.* **18** (1966), pp. 1091–1094.

[107] B. Bollobas, Three-graphs without two triples whose symmetric difference is contained in a third, *Discrete Math.* **8: 1** (1974), pp. 21–24.

[108] B. Bollobas, *Extremal graph theory*, Acad. Press, 1978. (Esp. Chapter III, Ex. 5.)

[109] B. Bollobas, *Combinatorics: Set systems, hypergraphs, families of vectors and combinatorial probability*, Cambridge Univ. Press, 1986.

[110] B. Bollobas, *Graph theory: an introductory course*, Springer, 1979.

[111] J.A. Bondy, U.S.R. Murthy, *Graph theory with applications*, North-Holland, 1976.

[112] J.A. Bondy, M. Sinonovits, Cycles of even length in graphs, *J. Combin. Th. Ser. B* **16: 2** (1974), pp. 97–105.

[113] W.G. Brown, On graphs that do not contain a Thomsen graph, *Cand. Math. Bull.* **9** (1966), pp. 281–285.

[114] W.G. Brown, P. Erdös, M. Simonovits, Algorithmic solution of extremal digraph problems, *Trans. Amer. Math. Soc.* **292: 2** (1985), pp. 421–449.

[115] J.P. Burling, S.W. Reyner, Some lower bounds for the Ramsey numbers, *J. Combin. Th. Ser. B* **13: 2** (1972), pp. 168–169.

[116] J. Campbell, A note on an optimal-fit method for dynamic allocation of storage, *Comput. J.* **14: 1** (1971),

[117] A.K. Chandra, C.K. Wong, Worst-case analysis of a placement algorithm related to storage allocation, *SIAM J. Comput.* **4: 3** (1975), pp. 249–263.

[118] E.G. Coffman, J.Y.-T. Leung, Combinatorial analysis of an efficient algorithm for processor and storage allocation, *18th Ann. Symp. Foundations Comput. Sci.* (1977), Amer. Math. Soc., pp. 214–221.

[119] P. Delsarte, J.M. Goethals, J.J. Seidel, Spherical codes and design, *Geom. Dedicata* **6: 3** (1977), pp. 363–388.

[120] J. Denes, A.D. Keedwell, *Latin squares and their applications*, Akad. Kiado, Budapest, 1974.

[121] P.J. Denning, The working set model for program behavior, *Comm. ACM.* **11: 5** (1968), pp. 323–333.

[122] P.Erdös, T. Gallai, On maximal paths and circuits of graphs, *Acta Math. Acad. Sci. Hungar.* **10** (1959), pp. 337–356.

[123] P. Erdös, M. Simonovits, A limit theorem in graph theory, *Studia Sci. Math. Hungar.* **1: 1-2** (1966), pp. 51–57.

[124] P. Erdös, A. Renyi, V.T. Sos, On a problem of graph theory, *Studia Sci. Math. Hungar.* **1: 1-2** (1966), pp. 215–235.

[125] P. Erdös, *The art of counting*, MIT, Mass., 1973.

[126] P. Erdös, R.L. Graham, bookOld and new problems and results in combinatorial number theory Kunding, Geneve, 1980.

[127] P. Erdös, R.K. Guy, J.W. Moon, On refining partitions, *J. London Math. Soc.* **9(2): 4** (1975), pp. 565–570.

[128] P. Erdös, A. Meir, V.T. Sos, P. Turan, On some applications of graph theory I, *Discrete Math.* **2** (1970), pp. 207–228.

[129] P. Erdös, A. Meir, V.T. Sos, P. Turan, On some applications of graph theory II, *Studies in Pure Math.* (1971), Acad. Press, pp. 89–100.

[130] P. Erdös, A. Meir, V.T. Sos, P. Turan, On some applications of graph theory III, *Cand. Math. Bull.* **15: 1** (1972), pp. 27–32.

[131] P. Erdös, L. Moser, An extremal problem in graph theory, *J. Austral. Math. Soc.* **11** (1970), pp. 42–47.

[132] P. Erdös, M. Simonovits, Compactness results in extremal graph theory, *Combinatorica* **2: 3** (1982), pp. 275–288.

[133] W. Fernandez de la Vega, Bin packing can be solved within $1 + \epsilon$ in linear time, *Combinatorica* **1: 4** (1981), pp. 349–355.

[134] P.D. Finch, On the Möbius-functions of a non singular binary relation, *Bull. Austral. Math. Soc.* **3** (1970), pp. 155–162.

[135] P.A. Frakaszek, J.P. Considine, Reduction of storage fragmentation on direct access devices, *IBM J. Res. Develop.* **23: 2** (1979), pp. 140–148.

[136] P. Frankl, Z. Furedi, Exact solution of some Turan type problems, *J. Comb. Th. Ser. A* **45** (1987), pp. 226–262.

[137] N. Gastinel, *Linear numerical analysis*, Acad. Press, 1970.

[138] G. Girand, Mojoretien du nombre de Ramsey ternaire-bicolore en (4,4), *C.R. Acad. Sc. Paris Ser. A* **269: 15** (1969), pp. 620–622.

[139] G. Girand, Sur le probleme de Goodman pour le quadrangles et la mojoretien des nombres de Ramsey, *J. Comb. Th. Ser. B* **27: 3** (1979), pp. 237–253.

[140] M. Goldberg, E.G. Straus, Norm properties of C-numerical radii, *Lin. Alg. & Appl.* **24** (1979), pp. 113–131.

[141] M. Goldberg, E.G. Straus, Combinatorial inequalities, matrix norms, and generalized numerical radii, *General Ineq. II Ser. Numer. Math.* **47** (1980), pp. 37–46.

[142] A.W. Goodman, On the sets of acquaintances and strangers at any party, *Amer. Math. Monthly* **66: 9** (1959), pp. 778–783.

[143] R.L. Graham, B.L. Rothschild, J.H. Spencer, *Ramsey theory*, Wiley, 1980.

[144] J. Graver, J. Yackel, Some graph theoretic results associated with Ramsey's theorem, *J. Comb. Th.* **4** (1968), pp. 125–175.

[145] C. Greene, D. Kleitman, Proof techniques in the theory of finite sets, *Studies in Combinatories* (ed.: G.C. Rota) (1978), MAA, pp. 22–79.

[146] R.E. Greenwood, A.M. Gleason, Combinatorial relations and chromatic graphs, *Cand. J. Math.* **7: 1** (1955), pp. 1–7.

[147] C.M. Grinstead, S.M. Roberts, On the Rmasey numbers $R(3,8)$ and $R(3,9)$, *J. Comb. Th. Ser. B* **33: 1** (1982), pp. 27–51.

[148] R.K. Guy, *Unsolved problems in number theory*, Springer, 1981.

[149] D. Hanson, Sum-free sets and Ramsey numbers, *Discrete Math.* **14** (1976), pp. 57–61.

[150] D. Hanson, J. Hanson, Sum-free sets and Ramsey numbers II, *Discrete Math.* **20: 3** (1977), pp. 295–296.

[151] Ph. Hanlon, The incidence algebra of a group reduced partially ordered set, *Lecture Notes in Math.* **829** (1981), pp. 148–156.

[152] G.H. Hardy, E.M. Wright, *An introduction to the theory of numbers*, Clarendon Press, 1945.

[153] D.S. Hirschberg, A class of dynamic memory alocation algorithms, *Comm. ACM* **16: 10** (1973), pp. 615–618.

[154] D.S. Johnson, Fast algorithms for bin packing, *J. Comp. Systems Sci.* **8** (1974), pp. 272–314.

[155] D.S. Johnson, A. Demers, J.D. Ullman, M.R. Garey, R.L. Graham, Worst-case performance bounds for simple one-dimensional packing algorithms, *SIAM J. Comput.* **3: 4** (1974), pp. 299–325.

[156] J.G. Kalbfleisch, R.G. Stanton, On the maximal triangle-free edge chromatic graphs in three colors, *J. Comb. Th.* **5: 1** (1968), pp. 9–20.

[157] Gy. Katona, Extremal problems for hypergraphs, *Combinatorics. MC Tract* **56** (1974), pp. 13–42.

[158] Gy. Katona, Grafok, vektorok es valoszinusegszamitasi egyenlotlensegek, *Mat. Lapok* **1–2** (1969), pp. 123–127.

[159] Gy. Katona, How many sums of vectors can lie ina circle of radius $\sqrt{2}$, *Comb. Th. & Appl.* **2** (1970), North-Holland, pp. 687–694.

[160] D.I. Kleitman, Hypergraphic extremal properties, *Surveys in Combinatorics. London Math. Soc. Lecture Notes* **38** (1979), pp. 44–65.

[161] T. Kovari, V.T. Sos, P. Turan, On a problem of Zarankiewicz, *Colloq. Math.* **3** (1954), pp. 50–57.

[162] G.W. Leibniz, *Mathematische Schriften*, **2** Berlin, 1850.

[163] D.C. Lewis, A generalized Möbius inversion formula, *Bull. Amer. Math. Soc.* **78** (1972), pp. 558–561.

[164] B. Lindstrom, On two generalizations of classical Möbius function, *Preprint Math. Inst. Univ. Stockholm* **14** (1975),

[165] L. Lovasz, *Combinatorial problems and exercises*, Akad. Kiado, 1979.

[166] L. Lovasz, M.D. Plummer, *Matching theory*, Akad. Kiado, 1986.

[167] P.A. MacMahon, *Combinatorial analysis*, 1–2 Chelsea, reprint, 1960.

[168] W. Mantel, Wisk. Opgaven, *Wisk. Genootschap* **10** (1907), pp. 60.

[169] B.H. Margolin, R.P. Pormelee, M. Schatroff, Analysis of free-storage algorithms, *IBM Systm J.* **10: 4** (1971), pp. 283–304.

[170] R. Mathon, Lower bounds for Ramsey numbers and association schemes, *J. Comb. Th. Ser. B* **42: 1** (1987), pp. 122–127.

[171] A.M. Odzlyko, N.J.A. Sloane, New bound of the number of unit spheres that can touch a unit sphere in n dimensions, *J. Comb. Th. (A)* **26: 2** (1979), pp. 210–214.

[172] J.L. Peterson, Th.A. Normal, Buddy systems, *Comm. ACM* **20: 6** (1977), pp. 421–430.

[173] B. Randell, A note on storage fragmentation and program segmentation, *Comm. ACM* **12: 7** (1969), pp. 365–372.

[174] R. Redheffer, C. Smith, On a surprizing inequality of Goldberg and Straus, *Amer. Math. Monthly* **87: 5** (1980), pp. 387–390.

[175] R. Redheffer, C. Smith, The case $n = 2$ of the Goldberg—Straus inequality, *General Inequalities II. Int. Ser. Num. Math.* **47** (1980), pp. 47–51.

[176] J.M. Robson, A bounded storage algorithm for copying cyclic structures, *Comm. ACM* **20: 6** (1977), pp. 431–440.

[177] J.M. Robson, Worst-case fragmentation of first-fit and best-fit storage allocation strategies, *Comp. J.* **20: 3** (1979), pp. 242–244.

[178] G.C. Rota, *Finite operator calculus*, Acad. Press, 1975.

[179] G.C. Rota, On the foundations of combinatorial theory I, *Z. Wahrsch. Verw. Gebiete* **2** (1964), pp. 340–368.

[180] D.L. Russell, Internal fragmentation in a class of buddy systems, *SIAM J. Comput.* **6: 4** (1977), pp. 607–621.

[181] N. Sauer, The largset number of edges of a graph such that not more than g intersect in a point or no more than n are independent, *Comb. Math. & Appl.* (D.J.A. Welsh (ed.)) Acad. Press, 1971.

[182] M.R. Schroeder, *Number theory in science and communication*, Springer, 1984.

[183] J.E. Shore, On the external storage fragmentation produced by first-fit and best-fit allocation strategies, *Comm. ACM* **18: 8** (1975), pp. 433–440.

[184] V.T. Sos, On extremal problems in graph theory, *Comb. Structures & Appl.* Gordon & Breach, (1970), pp. 407–410.

[185] R.P. Stanley, Theory and applications of plane partitions. Part 1, *Studies in Appl. Math.* **2** (1972),

[186] B.S. Stechkin, On a surprising fact in extremal set theory, *J. Comb. Th. Ser. A* **29: 3** (1980), pp. 368–369.

[187] J.J. Sylvester, Math. questions with their solutions, *The Educational Times* **41** (1884), 21.

[188] P. Turan, Egy grafelmeleti szelsoertekfeladatrol, *Mat. Lapok* **49** (1941), pp. 436–453.

[189] P. Turan, Applications of graph theory to geometry and potential theory, *Comb. Structures & Appl.* Gordon & Breach, (1970), pp. 423–434.

[190] P. Turan, On some applications of graph theory to analysis, *Proc. Int. Conf. Constructive Th. Varna, 1970* (1972), pp. 351–358.

[191] K. Walker, An upper bound for the Ramsey number $M(5,4)$, *J. Comb. Th. Ser. A* **11: 1** (1971), pp. 1–10.

[192] L. Weisner, Abstract theory of inversion of finite sets, *Trans. Amer. Math. Soc.* **38: 3** (1935), pp. 474–484.

[193] R.J. Wilson, Analysis situs, *Graph Th. with Appl. to Algorithms and Computer Sci.* Wiley, (1985), pp. 789–800.

[194] R.J. Wilson, The Möbius function in combinatorial mathematics, *Comb. math. & Appl.* Acad. Press, (1971), pp. 315–333.

[195] R.J. Wilson, The Selberg sieve for a lattice, *Comb. Th. & Appl.* **3** North-Holland, (1970), pp. 1141–1149.

[196] H.S. Wilf, The Möbius function in combinatorial analysis and chromatic graph theory, *Proof Techniques in Graph Th.* Acad. Press, (1969), pp. 179–188.

[197] A. Björner, P.H. Edelman, G.M. Ziegler, Hyperplane arangements wit a lattice of regions, *Discrete Comput. Geom.* **5** (1990), pp. 263–288.

[198] T. Brylawski, The lattice of integer partitions, *Discrete Math.* **6** (1973), pp. 201–219.

[199] P. Erdös, J. Lehner, The distribution of the number of summands in the partitions of a positive integer, *Duke Math. J.* **8** (1941), pp. 335–345.

[200] R. Frucht, G.-C. Rota, La funcion de Möbius para particiones de un conjunto, *Revista Scientia* **122** (1963), pp. 111–115.

[201] I. Rival, M. Stanford, Algebraic aspects of partition lattices, in: N. White (ed.) *Matroid Theory*, Vol. 3 (1992).

[202] I.P. Fotino, Generalized convolution rings of arithmetic functions, *Pacific J. Math.* **61** (1975), pp. 103–116.

[203] P. Hall, A contribution to the theory of groups of prime-power order, *Proc. London Math. Soc. (2)* **36** (1933/4), pp. 29–95.

[204] A. Sidorenko, What we know and what we do not know about Turan numbers, (To appear).

[205] S. Selah, Primitive recursive bounds for van der Waerden numbers, *J. Amer. Math. Soc.* **1**]1988 pp. 683–697.

[206] S.P. Radziszowski, Small Ramsey numbers, *Dept. Comp. Sc. School of Comp. Sc. and Inform. Technol., Rochester Inst. Technology* **RIT-TR-93-009** (Febr. 1993),

[207] M.G. Kendall, P.A.P. Moran, *Geometrical probability*, Griffin, 1963.

Index of Symbols

Index

acyclic 27
addition of multisets 13
algorithm 42
algorithm, efficiency of an 43
algorithm, exponential 43
algorithm, polynomial 43
allocation 156
allocation, dynamic 156
allocation, static 156
anti-reflexivity 22
anti-symmetry 22
antichain 23
atom 23

Bachet problem 77, 93
Bachet–Mendeleev problem 78
Bell number 13
Bellean 12
Bertrand paradox 150
bijection 18
bijective correspondence 18
bijective map 18
bin-packing problem 59
binary function 27
binary operation 20
binary operation, associative 21
binary operation, commutative 21
binary operation, distributive 21
binomial coefficient 21
blocks 86
Boolean 10
Boolean of a multiset 15
bound of complete packability 65

C-order 88
chain 23
chain, finite, length of a 23
chromatic number 102
class NP 46
class P 46
clique of dimension k 40
co-atom 23
combinatorial optimization problems 41
combinatorial problem, complexity of a 43

complete bipartite graph 101
complete families of jobs 165
complete graph 10, 101
complete image 17, 18
complete pre-image 17
complexity of a combinatorial problem 43
composition 50
concatenation 86
concatenation refinement order 88
conjecture, Turan 103
conjugate partition 52
constant A 139
constant B 140
constant C 141
constant δ 141
contact number 147
convolution, Möbius-stable 29
convolution, stable 29
correct kernel 28
correspondence 16
correspondence, bijective 18
correspondence, n-place 16
counter service problem 98
counting problems 38
cover of a set 9
cover of a set, block of a 9
cycle 10, 19, 27
cycle, orbit of a 19

de Morgan laws 9
degree cycle 110
description of functional assignment 165
dichotomy 22
directed graph 11
Dirichlet principle 53
distributivity 9
Dobinski's formula 13
domain of definition 17
duration of working times 44

edge chromatic number 133
efficiency of an algorithm 43
eigenvalue 151

Other *Mathematics and Its Applications* titles of interest:

P.H. Sellers: *Combinatorial Complexes. A Mathematical Theory of Algorithms.*
1979, 200 pp.   ISBN 90-277-1000-7

P.M. Cohn: *Universal Algebra.* 1981, 432 pp.
                        ISBN 90-277-1213- 1 (hb), ISBN 90-277-1254-9 (pb,

J. Mockor: *Groups of Divisibility.* 1983, 192 pp.   ISBN 90-277-1539-4

A. Wwarynczyk: *Group Representations and Special Functions.* 1986, 704 pp.
                        ISBN 90-277-2294-3 (pb), ISBN 90-277-1269-7 (hb)

I. Bucur: *Selected Topics in Algebra and its Interrelations with Logic, Number
Theory and Algebraic Geometry.* 1984, 416 pp.   ISBN 90-277-1671-4

H. Walther: *Ten Applications of Graph Theory.* 1985, 264 pp.
                                                            ISBN 90-277-1599-8

L. Beran: *Orthomodular Lattices. Algebraic Approach.* 1985, 416 pp.
                                                            ISBN 90-277-1715-X

A. Pazman: *Foundations of Optimum Experimental Design.* 1986, 248 pp.
                                                            ISBN 90-277-1865-2

K. Wagner and G. Wechsung: *Computational Complexity.* 1986, 552 pp.
                                                            ISBN 90-277-2146-7

A.N. Philippou, G.E. Bergum and A.F. Horodam (eds.): *Fibonacci Numbers and
Their Applications.* 1986, 328 pp.   ISBN 90-277-2234-X

C. Nastasescu and F. van Oystaeyen: *Dimensions of Ring Theory.* 1987, 372 pp.
                                                            ISBN 90-277-2461-X

Shang-Ching Chou: *Mechanical Geometry Theorem Proving.* 1987, 376 pp.
                                                            ISBN 90-277-2650-7

D. Przeworska-Rolewicz: *Algebraic Analysis.* 1988, 640 pp.   ISBN 90-277-2443-1

C.T.J. Dodson: *Categories, Bundles and Spacetime Topology.* 1988, 264 pp.
                                                            ISBN 90-277-2771-6

V.D. Goppa: *Geometry and Codes.* 1988, 168 pp.   ISBN 90-277-2776-7

A.A. Markov and N.M. Nagorny: *The Theory of Algorithms.* 1988, 396 pp.
                                                            ISBN 90-277-2773-2

E. Kratzel: *Lattice Points.* 1989, 322 pp.   ISBN 90-277-2733-3

A.M.W. Glass and W.Ch. Holland (eds.): *Lattice-Ordered Groups. Advances and
Techniques.* 1989, 400 pp.   ISBN 0-7923-0116-1

N.E. Hurt: *Phase Retrieval and Zero Crossings: Mathematical Methods in Image
Reconstruction.* 1989, 320 pp.   ISBN 0-7923-0210-9

Du Dingzhu and Hu Guoding (eds.): *Combinatorics, Computing and Complexity.*
1989, 248 pp.   ISBN 0-7923-0308-3

Other *Mathematics and Its Applications* titles of interest:

A.Ya. Helemskii: *The Homology of Banach and Topological Algebras.* 1989, 356 pp.   ISBN 0-7923-0217-6

J. Martinez (ed.): *Ordered Algebraic Structures.* 1989, 304 pp.
                                                                 ISBN 0-7923-0489-6

V.I. Varshavsky: *Self-Timed Control of Concurrent Processes. The Design of Aperiodic Logical Circuits in Computers and Discrete Systems.* 1989, 428 pp.
                                                                 ISBN 0-7923-0525-6

E. Goles and S. Martinez: *Neural and Automata Networks. Dynamical Behavior and Applications.* 1990, 264 pp.   ISBN 0-7923-0632-5

A. Crumeyrolle: *Orthogonal and Symplectic Clifford Algebras. Spinor Structures.* 1990, 364 pp.   ISBN 0-7923-0541-8

S. Albeverio, Ph. Blanchard and D. Testard (eds.): *Stochastics, Algebra and Analysis in Classical and Quantum Dynamics.* 1990, 264 pp.   ISBN 0-7923-0637-6

G. Karpilovsky: *Symmetric and G-Algebras. With Applications to Group Representations.* 1990, 384 pp.   ISBN 0-7923-0761-5

J. Bosak: *Decomposition of Graphs.* 1990, 268 pp.   ISBN 0-7923-0747-X

J. Adamek and V. Trnkova: *Automata and Algebras in Categories.* 1990, 488 pp.
                                                                 ISBN 0-7923-0010-6

A.B. Venkov: *Spectral Theory of Automorphic Functions and Its Applications.* 1991, 280 pp.   ISBN 0-7923-0487-X

M.A. Tsfasman and S.G. Vladuts: *Algebraic Geometric Codes.* 1991, 668 pp.
                                                                 ISBN 0-7923-0727-5

H.J. Voss: *Cycles and Bridges in Graphs.* 1991, 288 pp.   ISBN 0-7923-0899-9

V.K. Kharchenko: *Automorphisms and Derivations of Associative Rings.* 1991, 386 pp.   ISBN 0-7923-1382-8

A.Yu. Olshanskii: *Geometry of Defining Relations in Groups.* 1991, 513 pp.
                                                                 ISBN 0-7923-1394-1

F. Brackx and D. Constales: *Computer Algebra with LISP and REDUCE. An Introduction to Computer-Aided Pure Mathematics.* 1992, 286 pp.
                                                                 ISBN 0-7923-1441-7

N.M. Korobov: *Exponential Sums and their Applications.* 1992, 210 pp.
                                                                 ISBN 0-7923-1647-9

D.G. Skordev: *Computability in Combinatory Spaces. An Algebraic Generalization of Abstract First Order Computability.* 1992, 320 pp.   ISBN 0-7923-1576-6

E. Goles and S. Martinez: *Statistical Physics, Automata Networks and Dynamical Systems.* 1992, 208 pp.   ISBN 0-7923-1595-2

Other *Mathematics and Its Applications* titles of interest:

M.A. Frumkin: *Systolic Computations.* 1992, 320 pp.   ISBN 0-7923-1708-4

J. Alajbegovic and J. Mockor: *Approximation Theorems in Commutative Algebra.*
1992, 330 pp.   ISBN 0-7923-1948-6

I.A. Faradzev, A.A. Ivanov, M.M. Klin and A.J. Woldar: *Investigations in Algebraic Theory of Combinatorial Objects.* 1993, 516 pp.   ISBN 0-7923-1927-3

I.E. Shparlinski: *Computational and Algorithmic Problems in Finite Fields.* 1992,
266 pp.   ISBN 0-7923-2057-3

P. Feinsilver and R. Schott: *Algebraic Structures and Operator Calculus. Vol. I.*
Representations and Probability Theory. 1993, 224 pp.   ISBN 0-7923-2116-2

A.G. Pinus: *Boolean Constructions in Universal Algebras.* 1993, 350 pp.
                                                            ISBN 0-7923-2117-0

V.V. Alexandrov and N.D. Gorsky: *Image Representation and Processing. A*
Recursive Approach. 1993, 200 pp.   ISBN 0-7923-2136-7

L.A. Bokut' and G.P. Kukin: *Algorithmic and Combinatorial Algebra.* 1994,
384 pp.   ISBN 0-7923-2313-0

Y. Bahturin: *Basic Structures of Modern Algebra.* 1993, 419 pp.
                                                            ISBN 0-7923-2459-5

R. Krichevsky: *Universal Compression and Retrieval.* 1994, 219 pp.
                                                            ISBN 0-7923-2672-5

A. Elduque and H.C. Myung: *Mutations of Alternative Algebras.* 1994, 226 pp.
                                                            ISBN 0-7923-2735-7

E. Goles and S. Martínez (eds.): *Cellular Automata, Dynamical Systems and*
Neural Networks. 1994, 189 pp.   ISBN 0-7923-2772-1

A.G. Kusraev and S.S. Kutateladze: *Nonstandard Methods of Analysis.* 1994,
444 pp.   ISBN 0-7923-2892-2

P. Feinsilver and R. Schott: *Algebraic Structures and Operator Calculus. Vol. II.*
Special Functions and Computer Science. 1994, 148 pp.   ISBN 0-7923-2921-X

V.M. Kopytov and N. Ya. Medvedev: *The Theory of Lattice-Ordered Groups.*
1994, 400 pp.   ISBN 0-7923-3169-9

H. Inassaridze: *Algebraic K-Theory.* 1995, 438 pp.   ISBN 0-7923-3185-0

C. Mortensen: *Inconsistent Mathematics.* 1995, 155 pp.   ISBN 0-7923-3186-9

R. Abłamowicz and P. Lounesto (eds.): *Clifford Algebras and Spinor Structures.* A
Special Volume Dedicated to the Memory of Albert Crumeyrolle (1919–1992).
1995, 421 pp.   ISBN 0-7923-3366-7

W. Bosma and A. van der Poorten (eds.), *Computational Algebra and Number*
Theory. 1995, 336 pp.   ISBN 0-7923-3501-5

Other *Mathematics and Its Applications* titles of interest: